PYTHON AUTOMATION MASTERY
FROM NOVICE TO PRO

4 BOOKS IN 1

BOOK 1
PYTHON AUTOMATION MASTERY: A BEGINNER'S GUIDE

BOOK 2
PYTHON AUTOMATION MASTERY: INTERMEDIATE TECHNIQUES

BOOK 3
PYTHON AUTOMATION MASTERY: ADVANCED STRATEGIES

BOOK 4
PYTHON AUTOMATION MASTERY: EXPERT-LEVEL SOLUTIONS

ROB BOTWRIGHT

Published by Rob Botwright
Library of Congress Cataloging-in-Publication Data
ISBN 978-1-83938-516-2
Cover design by Rizzo

Disclaimer

The contents of this book are based on extensive research and the best available historical sources. However, the author and publisher make no claims, promises, or guarantees about the accuracy, completeness, or adequacy of the information contained herein. The information in this book is provided on an "as is" basis, and the author and publisher disclaim any and all liability for any errors, omissions, or inaccuracies in the information or for any actions taken in reliance on such information.

The opinions and views expressed in this book are those of the author and do not necessarily reflect the official policy or position of any organization or individual mentioned in this book. Any reference to specific people, places, or events is intended only to provide historical context and is not intended to defame or malign any group, individual, or entity.

The information in this book is intended for educational and entertainment purposes only. It is not intended to be a substitute for professional advice or judgment. Readers are encouraged to conduct their own research and to seek professional advice where appropriate.

Every effort has been made to obtain necessary permissions and acknowledgments for all images and other copyrighted material used in this book. Any errors or omissions in this regard are unintentional, and the author and publisher will correct them in future editions.

TABLE OF CONTENTS – BOOK 1 - PYTHON AUTOMATION MASTERY: A BEGINNER'S GUIDE

TABLE OF CONTENTS – BOOK 2 - PYTHON AUTOMATION MASTERY: INTERMEDIATE TECHNIQUES

TABLE OF CONTENTS – BOOK 3 - PYTHON AUTOMATION MASTERY: ADVANCED STRATEGIES

TABLE OF CONTENTS – BOOK 4 - PYTHON AUTOMATION MASTERY: EXPERT-LEVEL SOLUTIONS

Introduction

Welcome to the world of Python Automation Mastery, a comprehensive book bundle that will take you on an extraordinary journey from a novice to a pro in the realm of automation. Across four meticulously crafted volumes, we will embark on an exploration of Python's capabilities to automate tasks, solve complex problems, and enhance productivity.

In today's fast-paced and technology-driven world, automation has become an indispensable skill. Python, with its simplicity and versatility, stands out as the perfect tool for the job. Whether you're a complete beginner looking to enter the world of programming or an experienced developer aiming to sharpen your automation skills, this book bundle is your roadmap to mastering Python automation.

Book 1 - Python Automation Mastery: A Beginner's Guide
Our journey begins with Book 1, where we provide a solid foundation for automation. This volume is designed for those who are new to Python and programming in general. We will guide you through the fundamental concepts of Python and introduce you to the core principles of automation. By the end of this book, you'll have a strong grasp of Python basics and be ready to take your first steps in automating tasks that can simplify your everyday life.

Book 2 - Python Automation Mastery: Intermediate Techniques
Once you've established a solid foundation, we'll dive into intermediate-level techniques in Book 2. Here, we explore more complex automation challenges, such as web scraping, scripting, error handling, and data manipulation. You'll learn to tackle real-world automation tasks with confidence and precision. This book bridges the gap between beginners and those seeking to expand their automation skills further.

Book 3 - Python Automation Mastery: Advanced Strategies
In Book 3, we push the boundaries of automation by delving into advanced strategies. We introduce you to object-oriented programming, explore how to leverage external libraries and tools, and guide you through designing and implementing advanced automation projects. By the end of this volume, you'll be equipped to create automation

solutions that go beyond the basics and tackle complex, real-world challenges.

Book 4 - Python Automation Mastery: Expert-Level Solutions
The journey culminates in Book 4, where we explore expert-level solutions that will truly set you apart. Here, we tackle high-level use cases in domains like artificial intelligence, network security, and data analysis. You'll gain valuable insights into the most cutting-edge automation techniques and strategies, equipping you to handle the most demanding automation tasks with finesse.

Throughout this book bundle, we emphasize not only the "how" of automation but also the "why." We'll discuss best practices, ethical considerations, and the limitless possibilities of automation in our increasingly digital world. By the time you complete this journey, you won't just be a proficient Python programmer; you'll be an automation architect capable of innovating and creating solutions that can transform industries.

So, whether you're a curious beginner or an experienced developer seeking to enhance your automation prowess, join us on this exciting expedition from novice to pro in the world of Python automation. The Python Automation Mastery bundle is your passport to mastering the art of automation and harnessing the power of Python to simplify your life and supercharge your productivity.

BOOK 1
PYTHON AUTOMATION MASTERY
A BEGINNER'S GUIDE

ROB BOTWRIGHT

Chapter 1: The Python Automation Landscape

Automation is a fascinating and transformative concept that has taken the world by storm in recent years. It's a topic that's relevant to almost every aspect of our lives, from the devices we use at home to the technologies that power industries and businesses worldwide. As we delve into the world of automation, you'll discover how it simplifies tasks, enhances efficiency, and unlocks immense potential. So, let's embark on this journey together, where we'll explore the realm of automation and its mastery using Python.

Chapter 1: The Python Automation Landscape
In this chapter, we'll set the stage by discussing the landscape of automation. We'll explore the evolution of automation, from its humble beginnings to the sophisticated systems of today. You'll gain a deeper understanding of why automation is so crucial in today's fast-paced world and how Python plays a pivotal role in this landscape.

Chapter 2: Getting Started with Python
Before diving into automation, it's essential to establish a solid foundation in Python, a versatile and widely-used programming language. We'll walk you through the process of setting up your Python environment, ensuring you have all the tools you need to get started. From there, we'll guide you through the process of writing your first Python program, helping you grasp the basics and build confidence in your coding skills.

Chapter 3: Understanding Variables and Data Types

In this chapter, we'll delve into the fundamentals of Python programming. You'll learn about variables and data types, which are the building blocks of any code. We'll discuss how to declare variables, assign values to them, and work with various data types, such as integers, floats, strings, and more. By the end of this chapter, you'll have a solid grasp of how to handle data effectively in Python.

Chapter 4: Control Structures and Flow Control
Automation often involves making decisions and executing actions based on certain conditions. That's where control structures come into play. In this chapter, we'll explore conditional statements like if, else, and elif, which enable your programs to make intelligent choices. We'll also discuss looping constructs like for and while, allowing you to perform repetitive tasks efficiently.

Chapter 5: Functions and Modules in Python
As your Python skills grow, you'll find that functions and modules are essential for organizing and reusing code. In this chapter, we'll introduce you to the concept of functions, explaining how they encapsulate logic and make your code more modular. We'll also dive into modules, which are Python's way of organizing functions and variables into reusable packages.

Chapter 6: File Handling and Data Input/Output
Now that you've got a solid grasp of Python's fundamentals, it's time to explore how to work with external data sources. In this chapter, we'll cover file handling, which enables you to read and write data to files on your computer. You'll discover various techniques for reading and writing text and binary files, a skill that's invaluable in many automation tasks.

Chapter 7: Introduction to Web Scraping
The web is a vast source of data and information, and Python can be a powerful tool for extracting and manipulating this data. In this chapter, we'll introduce you to the world of web scraping, where you'll learn how to retrieve information from websites. You'll explore the ethics of web scraping and get hands-on experience using Python libraries to scrape data from web pages.

Chapter 8: Automating Tasks with Python Scripts
Automation often involves automating routine tasks that you perform regularly. In this chapter, we'll dive into the world of Python scripting. You'll discover how to create scripts that can automate tasks such as file management, data processing, and more. These scripts will save you time and effort while increasing your productivity.

Chapter 9: Error Handling and Debugging
No programming journey is complete without encountering errors. In this chapter, we'll teach you how to handle errors gracefully in your Python code. You'll learn about exceptions, which are Python's way of dealing with unexpected situations, and explore debugging techniques to identify and fix issues in your code effectively.

Chapter 10: Practical Projects for Beginners
To solidify your knowledge and put your skills to the test, we'll conclude our beginner's guide with practical projects. These projects are designed to challenge you and provide hands-on experience in automating real-world tasks. You'll have the opportunity to apply everything you've learned throughout the book and gain the confidence to tackle automation projects on your own.

Throughout this journey, you'll gain a deep appreciation for the power of automation and how Python can be your faithful companion in mastering it. So, let's dive in, explore, and unlock the limitless possibilities that automation with Python offers.

Python, often described as a versatile and user-friendly programming language, has firmly established itself as a vital tool in the world of automation. With its elegant syntax and a wealth of libraries, Python serves as an ideal choice for automating a wide array of tasks.

When it comes to automation, Python's role extends far and wide, from simplifying everyday chores on your computer to orchestrating complex processes in industries and enterprises. Its popularity in automation is not just a matter of chance; it's the result of its inherent qualities.

Python's readability and ease of use are undeniable strengths, making it accessible even to beginners in the world of programming. The simplicity of its syntax allows programmers to focus on solving problems rather than grappling with the intricacies of the language itself.
Moreover, Python boasts a vast ecosystem of libraries and frameworks tailored to different automation needs. Whether you're scraping data from websites, automating repetitive office tasks, or managing a fleet of servers, Python has a library or framework that simplifies and streamlines the process.

Python's popularity extends to the field of data analysis and manipulation. Data is the lifeblood of automation, and Python offers powerful libraries like Pandas and NumPy for handling, processing, and analyzing data.

In addition to data analysis, Python shines in the realm of data visualization. With libraries such as Matplotlib and Seaborn, you can effortlessly create insightful charts and graphs to convey your data-driven insights effectively.

Python's extensive support for web development also plays a significant role in automation. Through frameworks like Django and Flask, you can build web applications and APIs that automate various aspects of online interactions.

Python's adaptability is another key factor in its prominence in automation. It's equally proficient in the realms of Windows, macOS, and Linux, making it a versatile choice for automating tasks across different platforms.

Furthermore, Python's cross-platform compatibility extends to its ability to interface with a wide range of software and hardware systems. It can communicate with databases, network devices, IoT sensors, and more, enabling you to orchestrate diverse technologies seamlessly.

The open-source nature of Python is yet another reason behind its widespread adoption in automation. The Python community is vibrant and continually contributes to the language's growth. You can access a treasure trove of free and open-source libraries, tools, and resources that aid automation projects without incurring excessive costs.

Python's strong support for parallelism and concurrency is a game-changer in automation scenarios that require high performance and efficiency. Through libraries like asyncio and threading, Python empowers you to tackle tasks concurrently, making the most of modern multi-core processors.

Cybersecurity and ethical hacking are areas where Python's versatility shines even brighter. Security professionals and ethical hackers frequently use Python for tasks like penetration testing, vulnerability scanning, and network monitoring. Its extensive libraries, such as Scapy and PyCrypto, simplify these complex tasks.

Machine learning and artificial intelligence are at the forefront of automation innovation. Python's well-established libraries like TensorFlow, PyTorch, and scikit-learn make it an ideal choice for developing machine learning models that can automate decision-making processes, predict outcomes, and optimize operations.

When it comes to cloud computing and containerization, Python doesn't lag behind. It can integrate seamlessly with cloud platforms like AWS, Azure, and Google Cloud, enabling you to automate resource provisioning, scaling, and management. In the realm of containers, Python supports tools like Docker and Kubernetes, streamlining container orchestration and deployment.

Python's ecosystem of testing frameworks simplifies the crucial task of verifying the correctness and reliability of automation code. Whether it's unit testing, integration testing, or end-to-end testing, Python provides the tools and frameworks to ensure your automation solutions perform flawlessly.

In summary, Python's role in automation is multifaceted and continually expanding. Its simplicity, versatility, and extensive ecosystem of libraries and frameworks make it the go-to language for automation projects of all scales and complexities. Whether you're a beginner or an experienced programmer, Python has something to offer in your journey to master automation.

Chapter 2: Getting Started with Python

Let's begin our journey into the world of Python by first addressing a fundamental aspect: setting up your Python environment. You might be wondering, "Why is this important?" Well, my friend, having a properly configured Python environment is akin to having a well-equipped workspace before embarking on a creative project. It provides you with the essential tools and resources to write, test, and execute Python code effectively.

The first step in this process is to download Python itself. You can obtain the latest version of Python from the official Python website. Be sure to choose the appropriate version for your operating system, whether it's Windows, macOS, or Linux. Once you've downloaded the installer, run it and follow the installation instructions. Python's installation process is quite user-friendly and straightforward.

After Python is installed on your computer, it's time to verify that everything is set up correctly. Open your command prompt or terminal and type **python --version**. This command will display the installed Python version, confirming that Python is accessible from your command line. If you see the version number without any errors, you're on the right track.

The next step is to install a code editor or integrated development environment (IDE). A code editor is like your trusty notebook; it's where you'll write and edit your Python code. There are several options available, both free and paid, catering to different preferences and needs. Popular choices include Visual Studio Code, PyCharm, and Jupyter Notebook. Install the one that suits your workflow and style.

Once you have your code editor or IDE installed, you can open it and create a new Python file. This is where you'll write your Python code. Save the file with a **.py** extension, which indicates that it's a Python script. You're now all set to start coding!

It's important to note that Python comes with a built-in interactive interpreter known as the Python shell. You can access it by typing **python** in your command prompt or terminal. This is a handy way to quickly experiment with Python code and test small snippets without the need for a separate file.

Now, let's discuss virtual environments. A virtual environment is like having a separate workspace for each of your Python projects. It allows you to isolate project-specific dependencies and configurations, preventing conflicts between different projects. To create a virtual environment, use the **venv** module, which comes bundled with Python. Simply navigate to your project's directory in the command prompt or terminal and run **python -m venv venv_name**, replacing **venv_name** with your chosen name for the virtual environment.

To activate the virtual environment, use the following command:

On Windows: **venv_name\Scripts\activate**

On macOS and Linux: **source venv_name/bin/activate**

When the virtual environment is activated, you'll notice that your command prompt or terminal prompt changes, indicating the active virtual environment. You can now install project-specific packages without affecting your system-wide Python installation.

Installing packages is a common task in Python development. Python's package manager, **pip**, simplifies this process. To install a package, use the command **pip install**

package_name. For example, if you want to install the popular NumPy package for numerical computing, you can run **pip install numpy**. Pip will fetch the package and its dependencies from the Python Package Index (PyPI) and install them into your virtual environment.

Managing dependencies is crucial, especially when working on larger projects. To keep track of the packages your project depends on, you can create a **requirements.txt** file that lists all the packages and their versions. You can generate this file automatically by running **pip freeze > requirements.txt**. When sharing your project with others, they can recreate the same virtual environment by running **pip install -r requirements.txt**.

Now that your Python environment is up and running, you're ready to dive into the world of Python programming. With a well-configured environment, you can confidently write, test, and run Python code for a wide range of applications, from web development to data analysis and machine learning. Your journey into the exciting realm of Python has just begun, and there's a world of possibilities awaiting your exploration. Happy coding!

Congratulations! You've taken your first steps into the world of Python, and now it's time to write your very first Python program. Don't worry if you're feeling a bit nervous; everyone starts somewhere, and Python is known for its beginner-friendly nature.

Let's start by opening your chosen code editor or integrated development environment (IDE). Remember, this is where you'll be writing your Python code, so take a moment to familiarize yourself with the interface.

Once your code editor is open and ready, go ahead and create a new Python file. You can do this by selecting "New File" or a similar option in the menu.

Now, it's time to write your first line of Python code. In Python, the simplest program you can create is one that displays a message. To do this, type **print("Hello, World!")** in your Python file.

What you've just written is a Python statement that tells the computer to print the text "Hello, World!" to the screen. This is a traditional first program for many programmers, and it's a great way to get started.

Before running your program, take a moment to save your file. Choose a location on your computer where you'd like to save your Python program, and give it a meaningful name with the **.py** extension. For example, you can name it **hello_world.py**.

Now that your program is saved, you're ready to run it. In most code editors or IDEs, there's a "Run" or "Execute" button you can click. Alternatively, you can run the program from the command line by navigating to the directory where your program is saved and typing **python hello_world.py** (replace **hello_world.py** with the name of your Python file if it's different).

When you run the program, you should see the output on the screen: "Hello, World!" Congratulations! You've just written and executed your very first Python program.

Now, let's break down what you've done. In your program, **print()** is a Python function that takes an argument, in this case, the text you want to display. The text is enclosed in double quotation marks (") to indicate that it's a string, which is a sequence of characters.

The parentheses **()** after **print** are used to pass arguments to the function. In this case, you've passed the string **"Hello, World!"** as the argument, and the **print()** function displays that text on the screen.

Python is a dynamically-typed language, which means you don't need to declare the data type of a variable explicitly. Python determines the data type based on the value assigned to the variable. For example, you can create a variable **message** and assign the string **"Hello, Python!"** to it without specifying its data type.

Now, instead of directly passing the string to the **print()** function, you can use your **message** variable. Modify your program to look like this:

pythonCopy code

```
message = "Hello, Python!"  print(message)
```

When you run this program, you'll get the same output: "Hello, Python!" The variable **message** holds the string, and the **print()** function displays the contents of the variable.

Congratulations, you've just learned how to declare variables and use them in your Python programs! Variables are essential for storing and managing data in your code.

In Python, you can perform various operations with strings. For example, you can concatenate (combine) strings using the + operator. Modify your program to create a new message by concatenating two strings:

pythonCopy code

```
greeting = "Hello, "  subject = "Python!"  message = greeting + subject  print(message)
```

When you run this program, you'll see the output: "Hello, Python!" The + operator here combines the two strings to create the **message** variable.

In addition to concatenation, you can also perform other operations on strings, such as slicing, indexing, and formatting. Python provides a rich set of tools for working with text.

Now, let's explore another fundamental concept in programming: comments. Comments are explanatory notes within your code that are ignored by the computer when the program runs. They are intended for humans to read and understand the code better.

In Python, you can create comments using the # symbol. Anything following a # on a line is treated as a comment and is not executed as code. Comments are incredibly useful for documenting your code and explaining its logic to yourself and others.

For example:

pythonCopy code

```
# This is a comment  print("This is not a comment")
```

In this program, the **# This is a comment** line is a comment, while the **print("This is not a comment")** line is an actual Python statement.

Using comments in your code is a good practice that helps make your code more understandable and maintainable, especially when your programs become more complex.

Now that you've learned about variables, strings, and comments, you're well on your way to becoming a Python programmer. These concepts are the building blocks of Python and will be used extensively in your coding journey. So, keep practicing, experimenting, and exploring the exciting world of Python!

Chapter 3: Understanding Variables and Data Types

Now that you've embarked on your Python journey and written your first program, it's time to delve deeper into a fundamental concept: variables and data types.

In Python, variables are like containers that hold data. You can think of them as labels or names that you give to pieces of information. These labels make it easier to work with and manipulate data in your programs.

When you create a variable in Python, you're essentially reserving a spot in memory to store a value. This value can be a number, a piece of text, or even more complex data structures like lists or dictionaries.

Python is a dynamically-typed language, which means you don't need to declare the data type of a variable explicitly. Python determines the data type based on the value assigned to the variable.

For example, you can create a variable **age** and assign the integer **25** to it without specifying its data type. Python knows that **25** is an integer, so it assigns the integer data type to the variable **age**.

Similarly, you can create a variable **name** and assign the string **"Alice"** to it. Again, Python automatically recognizes that **"Alice"** is a string and assigns the string data type to the variable **name**.

Let's dive deeper into some common data types in Python:

Integers (int): Integers are whole numbers without a decimal point. You can use them to represent quantities like counts or ages. For example, **age = 25** assigns the integer **25** to the variable **age**.

Floats (float): Floats are numbers that can have a decimal point. They're used for representing measurements and other values with fractional parts. For example, **height = 1.75** assigns the float **1.75** to the variable **height**.

Strings (str): Strings are sequences of characters, like words or sentences. You can create strings by enclosing text in single or double quotation marks. For example, **name = "Alice"** assigns the string **"Alice"** to the variable **name**.

Booleans (bool): Booleans represent either true or false values. They're often used for making decisions in your code. For example, **is_student = True** assigns the boolean **True** to the variable **is_student**.

Lists: Lists are ordered collections of items. They can contain elements of different data types, and you can modify them. For example, **fruits = ["apple", "banana", "cherry"]** creates a list of strings and assigns it to the variable **fruits**.

Tuples: Tuples are similar to lists but are immutable, which means you can't change their contents once defined. They're often used when you want to create a collection of items that should not be modified. For example, **coordinates = (3, 4)** creates a tuple of two integers.

Dictionaries: Dictionaries are collections of key-value pairs. They're used for mapping one value (the key) to another (the value). For example, **person = {"name": "Alice", "age": 25}** creates a dictionary with string keys and values of various data types.

Now, let's talk about variable names. When naming your variables, there are some rules and conventions to keep in mind:

Variable names must start with a letter (a-z, A-Z) or an underscore (_).

They can contain letters, numbers, and underscores.

Variable names are case-sensitive, so **age** and **Age** are considered different variables.

Choose descriptive variable names that convey the purpose of the variable. For example, **total_cost** is more informative than **tc**.

It's also a good practice to follow naming conventions. For example, use lowercase letters and underscores for variable names (e.g., **total_cost**), and use uppercase letters for constants (e.g., **TAX_RATE**).

Now that you understand variables and data types, let's explore some basic operations you can perform with them. In Python, you can use operators to manipulate variables and data.

Arithmetic Operators: These operators allow you to perform mathematical calculations like addition, subtraction, multiplication, and division. For example, **result = 5 + 3** assigns the value **8** to the variable **result**.

Comparison Operators: These operators compare two values and return a boolean result (True or False). Examples include == for equality, != for inequality, < for less than, and > for greater than.

Logical Operators: Logical operators are used to combine or modify boolean values. They include **and, or,** and **not**. For example, **is_adult = age >= 18 and has_id_card** assigns **True** or **False** to the variable **is_adult** based on the conditions.

String Concatenation: You can use the + operator to concatenate (combine) strings. For example, **greeting = "Hello, " + name** combines the strings **"Hello, "** and **"Alice"** into a single string.

List and Tuple Operations: Lists and tuples support various operations like indexing (accessing elements by position), slicing (extracting portions), and appending (adding elements). For example, **first_fruit = fruits[0]** assigns the first item in the **fruits** list to the variable **first_fruit**.

Dictionary Operations: Dictionaries allow you to access values using keys. For example, **person_age = person["age"]** assigns the value **25** to the variable **person_age** by looking up the key **"age"** in the **person** dictionary.

As you gain more experience with Python, you'll discover many more operations and techniques for working with variables and data types. These foundational concepts are just the beginning of your Python journey. So, keep exploring, experimenting, and building your coding skills!

Welcome to the next chapter of your Python journey, where we'll dive into the fascinating world of working with strings and numbers. Strings and numbers are fundamental data types in Python, and you'll use them extensively in your coding adventures.

Let's begin with strings, which are sequences of characters, such as words or sentences. In Python, you can create strings by enclosing text in single (') or double (") quotation marks. For example, **name = "Alice"** or **message = 'Hello, World!'**.

Strings are versatile and offer a wide range of operations. You can concatenate (combine) strings using the + operator, like **greeting = "Hello, " + name**. This creates a new string that contains both the greeting and the name.

Another handy operation is string interpolation, where you embed variables or expressions within a string. Python provides two methods for string interpolation: f-strings and the **.format()** method. With f-strings, you can create dynamic strings by placing an **f** or **F** before the string and using curly braces {} to insert variables or expressions. For example, **message = f"Hello, {name}"**.

The **.format()** method allows you to insert variables or expressions into a string using placeholders enclosed in curly braces {}. You can then use the **.format()** method to replace the placeholders with values. For example, **message = "Hello, {}".format(name)**.

String slicing is another useful technique. It allows you to extract specific portions of a string. For example, you can use **substring = message[0:5]** to extract the first five characters from the **message** string. Python uses zero-based indexing, so **[0]** refers to the first character.

You can also manipulate strings by converting them to uppercase or lowercase using the **.upper()** and **.lower()** methods. For instance, **uppercase_name = name.upper()** would result in the string **"ALICE"**.

Python provides various string methods for performing tasks like finding substrings, replacing text, and checking if a string starts or ends with a specific sequence. For example, **contains = message.startswith("Hello")** would return **True** if the **message** string starts with "Hello."

Now, let's turn our attention to numbers. In Python, you can work with two primary numeric types: integers (**int**) and floating-point numbers (**float**). Integers are whole numbers without a decimal point, such as **5** or **-42**. Floating-point numbers, on the other hand, include decimal points, like **3.14** or **-0.5**.

Python provides various operators for performing arithmetic operations with numbers. You can use the **+** operator for addition, **-** for subtraction, ***** for multiplication, and **/** for division. For example, **result = 5 + 3** assigns the value **8** to the variable **result**.

Division in Python can yield different results depending on the types of numbers involved. When you divide two integers, Python returns a floating-point result if there's a remainder. For example, **result = 7 / 2** would result in **3.5**.

To perform integer division (floor division), you can use the **//** operator. It truncates the fractional part and returns an integer result. For example, **result = 7 // 2** would result in **3**.

Python also supports the modulo operator **%**, which gives you the remainder of a division operation. For instance, **remainder = 7 % 3** assigns the value **1** to the variable **remainder**.

Exponents can be calculated using the ****** operator. For example, **result = 2 ** 3** assigns the value **8** to the variable **result**.

Now, let's talk about a crucial concept when working with numbers: variable assignments and updates. You can assign a new value to a variable by using the **=** operator. For example, **x = 5** assigns the value **5** to the variable **x**.

Python also provides shortcuts for updating variables. For instance, the **+=** operator can be used to increment a variable's value. If **x** is **5**, then **x += 3** would result in **x** having the value **8**.

Similarly, you can use other operators like **-=** for subtraction, ***=** for multiplication, and **/=** for division to update variables more efficiently.

When working with numbers, it's essential to be mindful of data types. Python automatically converts between data types as needed. For example, if you add an integer and a float, Python promotes the integer to a float to perform the operation. However, you can explicitly convert data types using functions like **int()**, **float()**, and **str()**. For instance, **integer = int(3.14)** converts the floating-point number **3.14** to an integer.

Handling user input is another crucial aspect of working with both strings and numbers. You can use the **input()** function to capture user input as a string. For example, **name =**

input("Enter your name: ") would prompt the user to enter their name and store it as a string in the **name** variable.

To work with user input as numbers, you'll need to convert the input using functions like **int()** or **float()**. For example, **age = int(input("Enter your age: "))** captures the user's input as a string, converts it to an integer, and stores it in the **age** variable.

Finally, let's touch on some common tasks that involve both strings and numbers. Formatting output is an important skill, and Python provides various methods for this. You can use f-strings to create formatted strings with variables and expressions. For instance, **formatted_message = f"Hello, {name}! Your age is {age}."** combines strings and variables in a human-readable format.

Additionally, you can use string methods like **.format()** to create formatted strings. For example, **formatted_message = "Hello, {}. Your age is {}.".format(name, age)** creates a similar formatted message.

In this chapter, you've explored the fascinating world of strings and numbers in Python. These fundamental data types are the building blocks of many Python programs, and mastering them is a significant step in your journey as a Python developer. As you continue to learn and practice, you'll discover even more powerful ways to manipulate and work with strings and numbers in your Python projects. So, keep coding and exploring, my fellow Python enthusiast!

Chapter 4: Control Structures and Flow Control

Welcome to the realm of conditional statements in Python, a fundamental concept that allows your programs to make decisions and take different actions based on specific conditions. Think of conditional statements as the brains of your code, enabling it to respond intelligently to different scenarios.

At the heart of conditional statements are Boolean expressions. These are expressions that evaluate to either True or False. They serve as the foundation for making decisions in Python programs. For instance, you can create a Boolean expression like **age >= 18**, which checks whether a person's age is greater than or equal to 18.

Python provides a variety of conditional statements, with the most common one being the **if** statement. The **if** statement allows you to execute a block of code only if a specified condition is True. For example, you can use an **if** statement to check if a person is eligible to vote by evaluating the condition **if age >= 18:**.

You can extend the **if** statement by adding an **else** clause, which allows you to specify an alternative block of code to execute if the condition is False. This is incredibly useful for handling cases when the condition isn't met. For example, **if age >= 18: vote() else: print("You are not eligible to vote.")** provides a clear decision-making structure.

But what if you need to evaluate multiple conditions? Python offers the **elif** clause, which stands for "else if." You can use **elif** to specify additional conditions to check if the initial **if** condition is False. This allows you to handle multiple scenarios efficiently. For instance, you can use **if age < 18:**

print("You are not eligible to vote.") elif age == 18: print("You just turned 18! Congratulations!") else: vote() to cover various age scenarios.

Nested **if** statements are another powerful tool in your Python arsenal. They allow you to place one **if** statement inside another. This is particularly useful when you need to evaluate multiple conditions in a hierarchical manner. For example, you can use nested **if** statements to determine if a person is eligible to vote and, if so, check if they are registered to vote.

Logical operators play a vital role in crafting complex conditions. Python provides three main logical operators: **and**, **or**, and **not**. You can use **and** to combine multiple conditions, ensuring that all of them must be True for the entire condition to be True. Conversely, **or** allows you to specify that at least one of the conditions must be True for the entire condition to be True. The **not** operator negates a condition, making True conditions False and vice versa.

For example, you can use **if age >= 18 and is_registered:** to check if a person is both of voting age and registered to vote. Alternatively, you can use **if age < 18 or not is_registered:** to allow voting for those under 18 or those who aren't registered.

Conditional statements often involve comparing values. Python provides comparison operators like == (equal), != (not equal), < (less than), > (greater than), <= (less than or equal to), and >= (greater than or equal to) for this purpose. These operators allow you to compare variables, numbers, or other expressions.

For example, you can use **if age == 18:** to check if a person's age is exactly 18, or **if income >= 50000:** to determine if someone's income is greater than or equal to $50,000.

Conditional statements are not limited to handling single conditions. Python lets you create more complex decision-making structures by combining **if**, **elif**, and **else** clauses. This flexibility enables you to address a wide range of scenarios and make your code more robust.

The **if** statements we've discussed so far involve making decisions based on a single condition. However, there are cases where you need to evaluate multiple conditions simultaneously. Python provides the **if-elif-else** structure to handle such situations.

In this structure, you start with an **if** statement to check the first condition. If the first condition is True, the corresponding code block executes, and the program skips the remaining conditions. If the first condition is False, Python moves to the **elif** (else if) statement and evaluates its condition. This process continues until Python finds a True condition or reaches the **else** statement, which provides a default action if none of the conditions are met.

Let's take an example. Suppose you want to classify students' grades based on their scores. You can use the **if-elif-else** structure like this:

pythonCopy code

```
if score >= 90: grade = "A" elif score >= 80: grade = "B"
elif score >= 70: grade = "C" elif score >= 60: grade = "D"
else: grade = "F"
```

In this example, Python evaluates each condition in order. If the student's score is 95, for instance, the program will assign the grade "A" and skip the remaining conditions.

Conditional statements become even more versatile when you combine them with loops, functions, and data structures. They allow you to build complex programs that respond to changing conditions and user interactions.

In summary, conditional statements in Python are a fundamental tool for creating dynamic and intelligent programs. Whether you need to make simple decisions or handle complex scenarios, Python's **if**, **elif**, and **else** statements, along with logical and comparison operators, provide you with the building blocks to craft powerful and responsive code. As you continue to explore Python and write more code, you'll find that mastering conditional statements is a significant step toward becoming a proficient Python programmer. So, keep coding, experimenting, and building your programming skills!

Welcome to the fascinating world of looping and iteration in Python, where you'll learn how to perform repetitive tasks efficiently and effortlessly. Loops are like the workhorses of your code, allowing you to execute the same block of code multiple times, and they are an essential tool in any programmer's toolkit.

The most basic type of loop in Python is the **while** loop. It allows you to repeatedly execute a block of code as long as a specified condition remains True. Imagine a scenario where you want to count from 1 to 5:

pythonCopy code

```
count = 1  while count <= 5:  print(count) count += 1
```

In this example, the **while** loop keeps running as long as the condition **count <= 5** remains True. Inside the loop, we print the current value of **count**, and then increment it by 1 with **count += 1**. This process repeats until **count** becomes 6, at which point the condition becomes False, and the loop terminates.

But what if you want to perform an action a specific number of times? Enter the **for** loop. The **for** loop is excellent for iterating over a sequence of items, such as a list, tuple, or string, and performing the same action for each item. Here's an example of a **for** loop that iterates over a list of fruits:

pythonCopy code

```
fruits = ["apple", "banana", "cherry"] for fruit in fruits:
print(fruit)
```

In this code, the **for** loop iterates over each item in the **fruits** list and prints it. You can see how the loop simplifies repetitive tasks when you have a collection of items to process.

Python provides another powerful loop called the **for** loop with the **range()** function. This combination allows you to iterate over a sequence of numbers easily. For instance, if you want to print the numbers from 1 to 5, you can use the **range()** function like this:

pythonCopy code

```
for number in range(1, 6): print(number)
```

The **range(1, 6)** generates a sequence of numbers from 1 to 5 (inclusive), and the **for** loop iterates over these numbers one by one.

You can also use the **range()** function with a single argument to specify the end point. For example, **for number in range(6):** would accomplish the same task by starting from 0 and going up to, but not including, 6.

Sometimes, you may want to perform a specific action for each item in a sequence and keep track of the item's position or index. Python provides the **enumerate()** function for this purpose. Here's how you can use it:

pythonCopy code

```
fruits = ["apple", "banana", "cherry"] for index, fruit in
enumerate(fruits): print(f"Index {index}: {fruit}")
```

In this example, the **enumerate()** function pairs each item in the **fruits** list with its index, and the **for** loop iterates over these pairs. This allows you to access both the item and its index within the loop.

Python also offers the **break** and **continue** statements to control the flow of loops. The **break** statement allows you to exit a loop prematurely, even if the loop condition is still True. For example, you might use **break** to exit a loop when a certain condition is met:

pythonCopy code

```
numbers = [1, 2, 3, 4, 5, 6] for number in numbers: if number == 4: break print(number)
```

In this code, the **for** loop iterates over the **numbers** list. When it encounters the number 4, the **if** condition is met, and the **break** statement is executed, causing the loop to terminate.

On the other hand, the **continue** statement allows you to skip the rest of the current iteration and move to the next one. For example, you might use **continue** to skip even numbers when printing a list of numbers:

pythonCopy code

```
numbers = [1, 2, 3, 4, 5, 6] for number in numbers: if number % 2 == 0: continue print(number)
```

In this code, the **for** loop iterates over the **numbers** list. When it encounters an even number (determined by the **if** condition), the **continue** statement skips that iteration, and the loop proceeds to the next number.

Nested loops add another layer of complexity to your iteration techniques. They allow you to place one loop inside another, enabling you to work with two-dimensional data structures or perform more complex operations. For instance, you can use nested loops to iterate over a grid represented as a list of lists:

pythonCopy code

```
grid = [ [1, 2, 3], [4, 5, 6], [7, 8, 9] ] for row in grid: for
value in row: print(value)
```

In this example, the outer loop iterates over each row in the
grid, and the inner loop iterates over the values within each
row. This results in printing all the values in the grid.

Iterating over dictionaries in Python is also straightforward.
By default, a **for** loop iterates over the keys of a dictionary.
For example:

pythonCopy code

```
person = { "name": "Alice", "age": 30, "city":
"Wonderland" } for key in person: print(key, person[key])
```

In this code, the **for** loop iterates over the keys in the **person**
dictionary, and within the loop, you can access the
corresponding values using dictionary indexing.

If you want to iterate over both keys and values
simultaneously, you can use the **.items()** method of the
dictionary:

pythonCopy code

```
for key, value in person.items(): print(key, value)
```

This code iterates over both the keys and values of the
person dictionary, making it easy to work with dictionary
data.

In this exploration of looping and iteration techniques,
you've learned how to use **for** loops, **while** loops, and
specialized functions like **enumerate()** and **range()** to
perform repetitive tasks. You've also discovered how to
control the flow of loops using **break** and **continue**, as well
as how to work with nested loops and iterate over
dictionaries.

Chapter 5: Functions and Modules in Python

Welcome to the exciting world of functions in Python! Think of functions as reusable blocks of code that you can call by name to perform specific tasks. They are like mini-programs within your program, allowing you to break down complex problems into smaller, manageable parts.

Why use functions? Well, they offer several benefits. First and foremost, they promote code reusability. Instead of writing the same code multiple times, you can encapsulate it in a function and call that function whenever you need it. This not only saves you time but also makes your code more maintainable.

Creating a function in Python is a straightforward process. You start with the **def** keyword, followed by the function name and a pair of parentheses. Then, you add a colon to indicate the beginning of the function's code block. Here's a simple example of a function that greets you:

pythonCopy code

```
def greet(): print("Hello, there!")
```

In this example, we define a function named **greet()**, and its code block consists of a single line that prints "Hello, there!" when the function is called.

To use a function, you simply call it by name followed by parentheses. For example, to use our **greet()** function, you write **greet()** in your code. When you run the program, it will execute the code within the function, resulting in the greeting message being printed.

Functions can also accept parameters, which are values you pass to the function when you call it. These parameters allow your function to work with different data each time it's called. Let's modify our **greet()** function to accept a name as a parameter:

pythonCopy code

```python
def greet(name): print(f"Hello, {name}!")
```

Now, when you call **greet("Alice")**, it will print "Hello, Alice!" and if you call **greet("Bob")**, it will print "Hello, Bob!" This flexibility makes functions incredibly powerful for processing various inputs.

Functions can return values using the **return** statement. A function can perform calculations or processes and then return a result to be used elsewhere in your code. Here's an example of a function that adds two numbers and returns the result:

pythonCopy code

```python
def add(a, b): result = a + b return result
```

You can call this function like this: **sum = add(3, 5)**, and it will return **8**, which you can then assign to a variable named **sum**.

Python functions can also have default values for their parameters. This means that you can specify default values for some parameters, and if the caller doesn't provide values for those parameters, the defaults will be used. Here's an example:

pythonCopy code

```python
def greet(name, greeting="Hello"): print(f"{greeting}, {name}!") # You can call it with or without specifying the greeting greet("Alice") # Prints "Hello, Alice!" greet("Bob", "Hi") # Prints "Hi, Bob!"
```

In this code, the **greeting** parameter has a default value of "Hello," so if you call **greet("Alice")**, it uses the default greeting. However, if you provide a value for **greeting**, as in **greet("Bob", "Hi")**, it overrides the default value.

Functions can be organized into modules or files to make your code more modular and organized. You can create a Python file with functions and then import those functions into other Python files. This allows you to reuse your functions across different parts of your program or even in different projects.

For example, let's say you have a file named **math_operations.py** containing various math-related functions:

pythonCopy code

```
# math_operations.py def add(a, b): return a + b def
subtract(a, b): return a - b
```

You can then import these functions into another Python file:

pythonCopy code

```
# main.py from math_operations import add, subtract
result1 = add(3, 5) result2 = subtract(10, 4)
```

In this way, you can organize your code into manageable pieces, promoting code reuse and maintainability.

Another important concept in Python functions is scope. Scope refers to the region of your code where a variable is accessible. Variables defined inside a function have local scope, which means they are only accessible within that function. Variables defined outside of any function have global scope and can be accessed throughout your entire program.

Consider the following example:

pythonCopy code

```
x = 10  # Global variable  def my_function(): y = 5  # Local
variable  print(x)  # Accessing the global variable  print(y)  #
Accessing the local variable  my_function()  print(x)  #
Accessing the global variable outside the function  # print(y)
```

This would result in an error because y is local to the function

In this code, **x** is a global variable, so it can be accessed both inside and outside the **my_function()** function. However, **y** is a local variable defined within the function, so it can only be accessed from within the function.

Understanding scope is crucial for avoiding variable conflicts and ensuring that your code behaves as expected.

Recursion is a powerful technique in programming that involves a function calling itself. Recursive functions can be used to solve problems that have a repetitive or self-similar structure. One classic example of recursion is the computation of factorials.

Here's a recursive function that calculates the factorial of a number:

pythonCopy code

```
def factorial(n): if n == 0: return 1 else: return n * factorial(n - 1)
```

In this code, the **factorial()** function calls itself with a smaller argument (**n - 1**) until it reaches the base case, which is **n == 0**. At this point, the recursion stops, and the function starts returning values back up the call stack, ultimately computing the factorial.

Recursive functions can be elegant solutions to specific problems, but they require careful design to avoid infinite loops and excessive memory usage.

Another concept related to functions is lambda functions, also known as anonymous functions or lambda expressions. These are small, inline functions that are defined without a name. They are useful when you need a simple function for a short period, such as in sorting or filtering data.

Here's an example of a lambda function that adds two numbers:

pythonCopy code

```
add = lambda x, y: x + y result = add(3, 5)
```

Lambda functions are often used in conjunction with functions like **map()**, **filter()**, and **sorted()**. For instance, you can use a lambda function to sort a list of tuples based on the second element of each tuple:

pythonCopy code

```
data = [(1, 5), (3, 2), (2, 8)] sorted_data = sorted(data, key=lambda x: x[1])
```

In this code, the **key** parameter of the **sorted()** function accepts a lambda function that extracts the second element of each tuple for sorting.

As you delve deeper into Python, you'll encounter various built-in functions and libraries that offer a wide range of functionality, from mathematical operations to file handling and web interactions. These built-in functions, along with the functions you create, form the foundation of your Python programming toolkit.

In this exploration of creating and using functions, you've learned how functions can simplify your code, promote reusability, and enhance the organization of your programs. You've also discovered concepts like function parameters, return values, scope, recursion, and lambda functions that enable you to tackle a variety of programming tasks.

Functions are like building blocks that allow you to construct complex and powerful applications step by step. So, keep honing your function-writing skills, experiment with different techniques, and embrace the world of modular and reusable code!

Welcome to the world of organizing code with modules in Python! Modules are a fundamental concept that can help you manage the complexity of your programs and keep your code organized and readable. Think of modules as containers for Python code, allowing you to group related functions, classes, and variables together.

So, what exactly is a module? In simple terms, a module is a Python script or file containing Python code. This code can include function definitions, variable assignments, class definitions, or even executable code. Modules are essential for breaking down large programs into smaller, more manageable pieces.

To create a module, you start by writing your Python code in a separate **.py** file. Let's say you have a set of utility functions that you want to reuse across multiple projects. You can create a module named **my_utils.py** and define those functions inside it. Here's a basic example of a module containing two functions:

pythonCopy code

```
# my_utils.py def add(a, b): return a + b def subtract(a, b): return a - b
```

In this module, we've defined **add()** and **subtract()** functions that perform basic arithmetic operations. Now, you can reuse these functions in other Python scripts or modules by importing them.

To use functions from a module, you need to import the module into your Python script. You can do this using the **import** statement, followed by the name of the module (without the **.py** extension). Here's how you can import and use the functions from **my_utils.py**:

pythonCopy code

```
# main.py import my_utils result1 = my_utils.add(3, 5) result2 = my_utils.subtract(10, 4)
```

In this code, we import the **my_utils** module and then call its functions using the module name as a prefix (e.g., **my_utils.add()**). This helps prevent naming conflicts and allows you to organize your code more effectively.

Python also provides a way to import specific functions or variables from a module, rather than importing the entire module. This can make your code more concise and avoid cluttering your namespace with unnecessary names. To import specific functions or variables, you can use the **from** keyword followed by the module name and **import** statement. For example:

pythonCopy code

main.py from my_utils import add result = add(3 , 5)

In this example, we import only the **add()** function from the **my_utils** module. This allows us to use **add()** directly without the module prefix.

Modules can be organized into packages, which are directories that contain a collection of related modules. Packages help you create a structured hierarchy for your code, making it easier to manage and navigate. To create a package, you need to create a directory with a special **__init__.py** file (which can be empty) to indicate that the directory should be treated as a package. Here's an example of a simple package structure:

markdownCopy code

my_package/ __init__.py module1.py module2.py

In this structure, **my_package** is a package containing two modules, **module1.py** and **module2.py**. You can import modules from the package using dot notation. For instance, to import **module1** from **my_package**, you would write:

pythonCopy code

main.py from my_package import module1 result = module1.some_function()

This approach helps you organize your code into logical units and avoids naming clashes with other modules in different packages.

Python also includes a special module called **__init__.py** that can be used to specify which modules should be imported when the package itself is imported. This is useful for controlling the public interface of your package. For example, you can create an **__init__.py** file like this:

pythonCopy code

my_package/__init__.py from .module1 import some_function

In this **__init__.py** file, we specify that when the **my_package** package is imported, it should include the **some_function** from **module1**. This way, users of your package can access the functionality you want to expose without needing to know the internal structure of your package.

In addition to creating your own modules and packages, Python provides a vast standard library with modules for a wide range of tasks. These modules cover everything from working with files, handling data, and performing network operations to advanced topics like cryptography and web development. You can tap into this rich library to enhance your programs and save time by leveraging existing code.

When working on larger projects or collaborating with others, you might need to manage dependencies by using package managers such as **pip** and virtual environments like **virtualenv** or Python's built-in **venv**. These tools allow you to isolate project-specific dependencies, ensuring that your code runs consistently across different environments and systems.

In summary, organizing your code with modules and packages is a fundamental skill for any Python developer. Modules help you structure your code, promote code reuse, and manage the complexity of larger projects. Packages allow you to create organized hierarchies of modules, making your codebase more maintainable and navigable.

Whether you're creating your own modules and packages or utilizing the extensive Python standard library, understanding how to work with modules and packages is essential for building robust and maintainable Python applications. So, keep exploring, creating, and organizing your code using these powerful tools!

Chapter 6: File Handling and Data Input/Output

Welcome to the world of reading and writing files in Python, a fundamental skill for any programmer. Think of files as a way to interact with data that goes beyond what's stored in memory during program execution. Files allow you to read data from external sources, such as text files, databases, and web services, and write data for storage or sharing.

Let's start with the basics of reading files. To open and read a file in Python, you use the **open()** function, which takes two arguments: the file's name (including its path) and the mode in which you want to open it. The most common modes are "r" for reading, "w" for writing, and "a" for appending to an existing file.

For example, if you have a text file named "my_file.txt" in the same directory as your Python script, you can open it for reading like this:

pythonCopy code

```
file = open("my_file.txt", "r")
```

Once the file is open, you can read its contents using various methods. The simplest method is **read()**, which reads the entire contents of the file as a string. Here's an example:

pythonCopy code

```
file = open("my_file.txt", "r") contents = file.read()
print(contents) file.close()
```

In this code, we open the file, read its contents into the **contents** variable, print it, and then close the file using the **close()** method to free up system resources.

Another way to read a file is line by line using a **for** loop. This is useful for processing large files without loading the entire contents into memory. Here's how you can do it:

pythonCopy code

```
file = open("my_file.txt", "r") for line in file: print(line)
file.close()
```

In this example, we iterate through the file object with a **for** loop, and each iteration gives us one line from the file. This method is memory-efficient and works well with large files.

To make working with files more convenient and error-resistant, Python provides a **with** statement. It ensures that the file is properly closed when you're done with it, even if an error occurs within the block. Here's how you can read a file using the **with** statement:

pythonCopy code

```
with open("my_file.txt", "r") as file: contents = file.read()
print(contents)
```

With this approach, you don't need to explicitly close the file; Python takes care of it for you when you exit the **with** block.

Now, let's move on to writing files. To create a new file or overwrite an existing one, you can open it in write mode ("w"). If the file doesn't exist, Python will create it for you. Here's an example:

pythonCopy code

```
with open("new_file.txt", "w") as file: file.write("Hello, world!")
```

In this code, we open a file named "new_file.txt" in write mode and use the **write()** method to write the string "Hello, world!" into the file. If the file already exists, its previous contents will be replaced.

If you want to add content to an existing file without overwriting it, you can open the file in append mode ("a"). Here's how you can do it:

pythonCopy code

```
with    open("existing_file.txt",    "a")    as    file:
file.write("Appending some data.")
```

In this example, the text "Appending some data." is added to the end of the "existing_file.txt" without affecting its previous contents.

Working with binary files, such as images, audio, or video files, follows a similar process. You can open binary files using modes like "rb" (read binary) or "wb" (write binary). Here's an example of reading and writing a binary file:

pythonCopy code

```
# Reading a binary file with open("image.png", "rb") as
binary_file: data = binary_file.read() # Writing a binary file
with    open("copy_image.png",    "wb")    as    copy_file:
copy_file.write(data)
```

In this code, we first read the contents of the "image.png" file in binary mode and store it in the **data** variable. Then, we create a new binary file named "copy_image.png" and write the data into it, effectively creating a copy of the original image.

Python also provides a way to work with files as objects using the **io** module. This module offers classes like **TextIOWrapper** and **BytesIO** for reading and writing text and binary data, respectively. These classes provide additional functionalities and can be useful for advanced file operations.

When dealing with files, it's essential to handle exceptions that may occur, such as file not found errors or permission issues. Python's **try** and **except** statements allow you to gracefully handle these exceptions and provide meaningful error messages to users.

Here's an example of opening a file with error handling:
pythonCopy code

```
try: with open("nonexistent_file.txt", "r") as file: contents
= file.read() print(contents) except FileNotFoundError:
print("The file does not exist.") except PermissionError:
print("Permission to read the file is denied.")
```

In this code, we attempt to open a file that doesn't exist
("nonexistent_file.txt"). If the file is not found, a
FileNotFoundError exception is raised, and we handle it by
printing an error message.

In addition to basic file operations, Python offers libraries
and modules for more advanced file handling tasks, such as
working with CSV files (**csv** module), JSON data (**json**
module), and various data formats and databases.

File handling is a fundamental skill that every Python
developer should master. Whether you're reading
configuration files, processing data, or generating reports,
the ability to work with files efficiently and effectively is
crucial for building practical applications.

In summary, reading and writing files in Python is a vital skill
that opens up a world of possibilities for data manipulation
and interaction with external resources. From text and
binary files to error handling and advanced file formats,
understanding file operations is a fundamental step in your
journey as a Python programmer.

So, keep exploring and experimenting with file handling
techniques, and you'll find that your ability to work with data
and files will be an invaluable asset in your coding
endeavors!

Welcome to the exciting world of working with different file
formats in Python! As you continue your journey in
programming, you'll encounter a wide variety of file formats
used for storing and exchanging data. Understanding how to
work with these formats is an essential skill for any
developer.

Let's start by exploring one of the most common file formats: plain text files. Text files are versatile and easy to work with. They are used for everything from configuration files to data storage and program input/output.

In Python, reading and writing text files is a straightforward process. You can use the built-in **open()** function to open a text file, specify the "r" mode for reading or "w" mode for writing, and then manipulate the file's contents.

For example, suppose you have a text file named "data.txt" with the following content:

Copy code

John Doe Alice Johnson Bob Smith

You can read this file and print its contents like this:

pythonCopy code

```
with open("data.txt", "r") as file: contents = file.read()
print(contents)
```

This code opens "data.txt" in read mode, reads its content into the **contents** variable, and then prints it to the console. Text files are an excellent choice for storing human-readable data and configuration settings.

Moving on, let's explore another widely used format: Comma-Separated Values (CSV). CSV files are commonly used to store tabular data, such as spreadsheets or databases. Each line in a CSV file represents a row of data, and fields within each row are separated by commas (or other delimiters).

To work with CSV files in Python, you can use the **csv** module, which provides functions for reading and writing CSV data. Here's an example of reading data from a CSV file:

pythonCopy code

```
import csv with open("data.csv", "r", newline="") as file:
csv_reader = csv.reader(file) for row in csv_reader:
print(row)
```

In this code, we import the **csv** module, open "data.csv" in read mode, and use **csv.reader** to read its contents. The **newline=""** argument is used to ensure that newline characters are handled correctly. We then iterate through the rows and print each row as a list of values.

To write data to a CSV file, you can use **csv.writer**. Here's an example:

pythonCopy code

```
import csv data = [ ["Name", "Age"], ["Alice", 30], ["Bob", 25], ["Charlie", 35] ] with open("output.csv", "w", newline="") as file: csv_writer = csv.writer(file) csv_writer.writerows(data)
```

In this example, we create a list of lists called **data**, where each inner list represents a row of data. We then open "output.csv" in write mode and use **csv.writer** to write the data to the file. The **writerows()** method writes multiple rows at once.

Moving on to JSON (JavaScript Object Notation), which is a popular format for data interchange. JSON is a lightweight and human-readable format used for storing structured data. It's commonly used in web applications for sending and receiving data.

In Python, you can work with JSON using the built-in **json** module. You can convert Python data structures, such as dictionaries and lists, to JSON and vice versa. Here's an example of reading and writing JSON data:

pythonCopy code

```
import json # Reading JSON with open("data.json", "r") as file: data = json.load(file) print(data) # Writing JSON data = {"name": "Alice", "age": 30} with open("output.json", "w") as file: json.dump(data, file, indent=4)
```

In this code, we use **json.load()** to read JSON data from "data.json" and **json.dump()** to write JSON data to "output.json" with an indentation for readability.

Another common file format for data storage and interchange is XML (eXtensible Markup Language). XML is a flexible format used for representing structured data. It's commonly used in web services, configuration files, and data storage.

Python provides the **xml.etree.ElementTree** module for parsing and manipulating XML data. Here's an example of parsing an XML file:

pythonCopy code

```
import xml.etree.ElementTree as ET # Parsing XML tree = ET.parse("data.xml") root = tree.getroot() # Accessing XML elements for person in root.findall("person"): name = person.find("name").text age = person.find("age").text print(f"Name: {name}, Age: {age}")
```

In this code, we use **ET.parse()** to parse an XML file and obtain the root element. We then iterate through the "person" elements and access their child elements using the **find()** method.

To create and write XML data, you can use the same **xml.etree.ElementTree** module. Here's an example of creating and writing XML data:

pythonCopy code

```
import xml.etree.ElementTree as ET # Creating XML data root = ET.Element("people") person1 = ET.SubElement(root, "person") name1 = ET.SubElement(person1, "name") name1.text = "Alice" age1 = ET.SubElement(person1, "age") age1.text = "30" #
```

Writing XML tree = ET.ElementTree(root) tree.write("output.xml")

In this code, we create an XML structure using elements and subelements, and then write it to "output.xml" using the **write()** method.

Moving on to yet another file format, we have Excel spreadsheets. Excel files are widely used for data storage, analysis, and reporting. Python provides the **openpyxl** library for working with Excel files.

Here's an example of reading data from an Excel file:

pythonCopy code

```
import openpyxl # Reading Excel workbook = openpyxl.load_workbook("data.xlsx") sheet = workbook.active for row in sheet.iter_rows(values_only=True): print(row)
```

In this code, we use **openpyxl.load_workbook()** to load an Excel file, access the active sheet, and iterate through the rows to print their values.

To write data to an Excel file, you can create a new workbook and add data to it. Here's an example:

pythonCopy code

```
import openpyxl # Writing Excel workbook = openpyxl.Workbook() sheet = workbook.active data = [ ["Name", "Age"], ["Alice", 30], ["Bob", 25], ["Charlie", 35] ] for row in data: sheet.append(row) workbook.save("output.xlsx")
```

In this code, we create a new workbook, add data to the active sheet using **sheet.append()**, and then save it as "output.xlsx".

The last format we'll explore here is SQLite databases. SQLite is a lightweight and embedded relational database system that's commonly used in mobile apps and small-scale applications. Python provides the **sqlite3** module for working with SQLite databases.

Here's an example of creating and querying an SQLite database:

pythonCopy code

```
import sqlite3 # Creating a database and table conn = sqlite3.connect("my_database.db") cursor = conn.cursor() cursor.execute('''CREATE TABLE IF NOT EXISTS employees (id INTEGER PRIMARY KEY, name TEXT, age INTEGER)''') # Inserting data cursor.execute("INSERT INTO employees (name, age) VALUES (?, ?)", ("Alice", 30)) cursor.execute("INSERT INTO employees (name, age) VALUES (?, ?)", ("Bob", 25)) cursor.execute("INSERT INTO employees (name, age) VALUES (?, ?)", ("Charlie", 35)) # Querying data cursor.execute("SELECT * FROM employees") data = cursor.fetchall() for row in data: print(row) # Committing changes and closing the connection conn.commit() conn.close()
```

In this code, we connect to an SQLite database, create an "employees" table, insert data into it, and query the data. Finally, we commit the changes and close the database connection.

Working with different file formats is a valuable skill for any programmer. It enables you to handle a wide range of data sources and formats, making your programs more versatile and adaptable. Whether you're reading text files, parsing JSON, working with XML, managing Excel spreadsheets, or interacting with databases, Python provides libraries and modules to simplify these tasks.

Chapter 7: Introduction to Web Scraping

Welcome to the fascinating realm of web scraping, a powerful technique that allows you to extract data from websites and use it for various purposes. Web scraping is like having a digital pair of hands to collect information from the vast ocean of the internet.

Imagine you want to gather data from a website, such as news articles, product details, or weather information. Instead of manually copying and pasting this data, which would be time-consuming and error-prone, you can write a Python script to automate the process. This is where web scraping comes into play.

At its core, web scraping involves sending HTTP requests to a website, retrieving the HTML content of web pages, and then parsing and extracting the desired information. Python provides powerful libraries like **requests** for making HTTP requests and **Beautiful Soup** for parsing HTML.

Here's a simple example of how you can use Python to fetch and display the title of a web page:

pythonCopy code

```
import requests from bs4 import BeautifulSoup # Send an HTTP GET request response = requests.get("https://www.example.com") # Check if the request was successful if response.status_code == 200: # Parse the HTML content of the page soup = BeautifulSoup(response.text, "html.parser") # Find and print the page title title = soup.title.string print(f"The title of the page is: {title}") else: print("Failed to retrieve the web page.")
```

In this code, we import the **requests** library to send an HTTP GET request to a website (in this case, "https://www.example.com"). We check the response status code to ensure the request was successful (a status code of 200 means success). Then, we use **Beautiful Soup** to parse the HTML content of the page and extract the title, which we print to the console.

Web scraping can be incredibly valuable for a wide range of applications, from data analysis and research to creating customized datasets and monitoring websites for updates. However, it's essential to approach web scraping with ethics and legal considerations in mind.

Ethics and legality play a crucial role in web scraping because accessing websites and collecting data without permission can lead to various issues. Here are some key ethical and legal considerations when it comes to web scraping:

Respect Robots.txt: Many websites have a **robots.txt** file that tells web crawlers (like search engine bots) which parts of the site are off-limits. Always check a site's **robots.txt** file before scraping, and adhere to its rules.

Terms of Service: Review a website's terms of service or terms of use to see if web scraping is explicitly prohibited. Some websites explicitly forbid web scraping in their terms, while others may have specific guidelines for web scraping.

Public vs. Private Data: Distinguish between publicly available data and private or sensitive information. Collecting and using sensitive information without permission can lead to legal consequences.

Rate Limiting: Be mindful of the rate at which you make requests to a website. Rapid and excessive requests can put a strain on a website's server and may be seen as a denial-of-service attack. Use rate limiting to avoid overloading a site.

User-Agent Identification: When making HTTP requests, include a user-agent header that identifies your script or bot.

Providing a descriptive user-agent can help website administrators identify your intentions and contact you if necessary.

Consent and Transparency: If you plan to use scraped data for commercial purposes or distribute it, consider seeking permission from the website owner. Being transparent about your data collection practices can build trust.

Copyright and Attribution: Respect copyright laws when scraping content. Always attribute the source of data when using it, especially if it's copyrighted material.

Use Legal Sources: Ensure that the websites you scrape are legal sources of information. Avoid scraping websites that host illegal or copyrighted content.

Personal Data: Be cautious when handling personal data. Collecting and processing personal information is subject to strict privacy regulations in many countries.

It's important to note that web scraping laws and regulations can vary by country and jurisdiction. What may be legal in one place may not be in another. Therefore, it's crucial to research and understand the legal landscape in your specific context.

Many websites have adopted measures to protect themselves against web scraping, such as implementing CAPTCHA challenges or IP blocking. Attempting to circumvent these measures may violate both ethics and the law.

In summary, web scraping is a powerful tool for data collection and automation, but it comes with ethical and legal responsibilities. It's essential to approach web scraping with respect for website owners' rules and users' privacy, and to be aware of the legal implications in your jurisdiction. Responsible web scraping can unlock valuable data while maintaining a respectful and lawful presence on the web.

Let's dive deeper into the world of web scraping and explore the libraries and tools that can make your web scraping journey more efficient and enjoyable. As you embark on this exciting adventure, you'll find that having the right tools at your disposal can make a significant difference in your ability to extract data from the web.

Python, with its extensive ecosystem of libraries, is a fantastic choice for web scraping. While we've already introduced the basics of web scraping using the **requests** library and **Beautiful Soup**, there are several other libraries and tools that can enhance your web scraping capabilities.

One such library is **Scrapy**, a powerful and highly customizable web scraping framework. Unlike **Beautiful Soup**, which primarily focuses on parsing HTML and XML, **Scrapy** is a complete framework designed specifically for web crawling and data extraction. It allows you to define how you want to navigate websites, what data to extract, and how to store that data.

With **Scrapy**, you can create web scraping spiders, which are essentially scripts that navigate websites and extract data according to your specifications. These spiders can follow links, submit forms, and handle various complexities often encountered in web scraping tasks. **Scrapy** provides a convenient command-line tool for creating and running spiders, making it a popular choice for scraping large and complex websites.

To get started with **Scrapy**, you'll need to install it first using **pip**:

pythonCopy code

```
pip install scrapy
```

Once **Scrapy** is installed, you can create a new Scrapy project and define your spiders. Writing a Scrapy spider involves defining the website you want to scrape, specifying the data you want to extract, and implementing callback functions to process the extracted data.

Another valuable tool in the web scraping arsenal is **Selenium**. Unlike **Scrapy**, which is primarily used for static websites, **Selenium** is designed for scraping dynamic web pages, especially those that rely on JavaScript for content loading and interactivity.

Selenium allows you to automate a web browser (such as Chrome or Firefox) and interact with web pages just as a human user would. This means you can click buttons, fill out forms, and scroll through pages, enabling you to access data that may be hidden behind JavaScript-driven user interfaces.

To use **Selenium**, you'll need to install it along with a web driver for your chosen browser:

pythonCopy code

```
pip install selenium
```

Additionally, you'll need to download the appropriate web driver for your browser and set its path in your Python script. For example, if you're using Chrome, you would download the ChromeDriver and specify its path like this:

pythonCopy code

```
from selenium import webdriver driver = webdriver.Chrome(executable_path='/path/to/chromedriver')
```

Once **Selenium** is set up, you can automate your interactions with web pages, navigate through them, and extract data as needed. **Selenium** is particularly useful when dealing with websites that heavily rely on JavaScript for rendering content.

Moving on, another fantastic library for web scraping is **Pandas**, a versatile data manipulation library that can help you clean, transform, and analyze the data you've scraped. With **Pandas**, you can easily load scraped data into data frames, perform operations on it, and export it to various formats like CSV or Excel.

For example, after scraping data using **Beautiful Soup** or **Scrapy**, you can convert it into a **Pandas** data frame like this: pythonCopy code

```
import pandas as pd # Sample scraped data data = [ {"name": "Alice", "age": 30}, {"name": "Bob", "age": 25}, {"name": "Charlie", "age": 35}] # Convert to a Pandas data frame df = pd.DataFrame(data) # Perform data analysis and transformation mean_age = df["age"].mean() print(f"Mean age: {mean_age}")
```

In this example, we create a **Pandas** data frame from the scraped data and calculate the mean age. **Pandas** provides a wide range of data manipulation and analysis tools, making it an invaluable companion in your web scraping projects.

Next, let's explore **Requests-HTML**, a library that combines the simplicity of **requests** with the power of **Beautiful Soup**. It's an excellent choice for straightforward web scraping tasks where you need to make HTTP requests and extract data from the HTML content.

Requests-HTML allows you to make HTTP requests and parse HTML in a concise and Pythonic way. It provides features like CSS selector-based querying, making it easy to extract specific elements from web pages.

Here's an example of how you can use **Requests-HTML** to scrape the titles of articles from a website: pythonCopy code

```
from requests_html import HTMLSession # Create an HTML session session = HTMLSession() # Send an HTTP GET request response = session.get("https://example.com") # Use CSS selectors to extract data titles = response.html.find("h2") for title in titles: print(title.text)
```

In this code, we create an HTML session, send an HTTP GET request, and use CSS selectors to extract **h2** elements from the web page. **Requests-HTML** simplifies the process of making requests and parsing HTML content.

Lastly, it's crucial to mention **Splash**, a headless browser rendering service often used in conjunction with **Scrapy**. **Splash** allows you to render JavaScript-driven web pages within your web scraping process. This is particularly valuable when dealing with single-page applications (SPAs) and websites that load content dynamically using AJAX requests.

To use **Splash**, you typically run a **Splash** server and send HTTP requests to it, along with Lua scripts that specify how the web page should be rendered. **Scrapy** can be configured to work seamlessly with **Splash**, allowing you to scrape data from complex, JavaScript-heavy websites.

In this exploration of web scraping libraries and tools, we've covered a range of options, each suited to different web scraping scenarios. Whether you're dealing with static or dynamic websites, small or large-scale scraping tasks, Python offers libraries and tools to meet your needs.

However, it's essential to approach web scraping with responsibility and respect for the websites you're scraping. Always check a website's terms of service and **robots.txt** file, and be mindful of rate limiting and ethical considerations.

In your web scraping journey, you'll find that choosing the right combination of tools and libraries depends on the specific requirements of your project. With the knowledge of these tools and a sense of ethics, you'll be well-equipped to navigate the vast web and extract the data you need for your applications, research, or analysis. So, go forth and explore the wealth of information available on the web, responsibly and ethically.

Chapter 8: Automating Tasks with Python Scripts

Welcome to the exciting world of writing effective automation scripts! In this chapter, we'll explore the art and science of crafting scripts that not only get the job done but also do it efficiently, reliably, and maintainably. Writing automation scripts can be a rewarding experience, enabling you to automate repetitive tasks, streamline workflows, and boost productivity.

Before we dive into the technical aspects, let's start with the fundamental question: What is an automation script? At its core, an automation script is a program or script that performs tasks automatically, without human intervention. These tasks can range from simple file operations to complex data analysis, and everything in between. Automation scripts are your digital helpers, carrying out instructions to save you time and effort.

To write effective automation scripts, you need a clear understanding of the task or process you want to automate. Before you even start writing code, take some time to define the objectives, scope, and requirements of your automation project. What are you trying to achieve, and what are the specific steps involved? The better you understand the problem, the easier it will be to design a solution.

Next, consider the tools and technologies at your disposal. What programming language will you use for your script? The choice of programming language depends on your familiarity with it and its suitability for the task. Python, for example, is a versatile and beginner-friendly language commonly used for automation due to its readability and a rich ecosystem of libraries.

Once you've outlined your goals and chosen your tools, it's time to start coding. However, before you dive headfirst into writing code, remember that good scripts are often characterized by simplicity and clarity. Avoid the temptation to overcomplicate your code. Keep it clean, concise, and easy to understand. Remember that you might not be the only one maintaining the script, so make it as accessible as possible.

One essential aspect of writing effective automation scripts is error handling. Automation scripts should be robust and able to handle unexpected situations gracefully. This means anticipating potential errors and implementing error-handling mechanisms. For example, if your script is reading data from a file, consider what happens if the file doesn't exist or is empty. By adding error-handling code, you can prevent crashes and make your script more reliable.

Documentation is often an overlooked but critical aspect of scripting. A well-documented script is like a roadmap that helps you and others understand how the script works, its purpose, and how to use it. Don't skimp on comments and documentation. Explain your code's logic, provide usage examples, and include any prerequisites or dependencies. Good documentation makes it easier to maintain and troubleshoot your scripts in the long run.

Modularization is another powerful technique in script writing. Rather than writing a monolithic script, consider breaking it down into smaller, reusable functions or modules. Modularization promotes code reusability, readability, and easier debugging. It also allows you to tackle complex tasks one step at a time, making your scripts more manageable.

When writing automation scripts, consider the input and output formats. How will the script receive input, and how will it present output? Providing clear and user-friendly input and output options can significantly enhance the usability of your scripts. For example, if your script takes command-line arguments, ensure that the argument names and descriptions are well-documented.

Testing is a crucial part of the scripting process. Before deploying your automation script in a production environment, thoroughly test it in a controlled setting. Test various scenarios, including edge cases and unexpected inputs, to ensure the script behaves as expected. Automated testing frameworks can be valuable for automating the testing process itself.

Version control is your best friend when it comes to script development. Use a version control system like Git to track changes to your scripts over time. This allows you to collaborate with others, revert to previous versions if issues arise, and maintain a history of your script's development.

Now, let's talk about code organization. Effective scripts are organized logically, with a clear structure. You can follow established coding conventions or create your own, but consistency is key. Organize your code into functions, classes, and modules, and give meaningful names to variables and functions. This makes your code self-explanatory and easier to maintain.

Another important consideration is performance optimization. While it's essential to keep your code simple and readable, it's also important to make it efficient. Profile your code to identify bottlenecks and optimize critical sections. Techniques like caching, parallel processing, and algorithmic improvements can significantly boost script performance.

Security is a paramount concern in script development. If your script deals with sensitive data or interacts with external systems, take steps to secure it. Avoid hardcoding sensitive information like passwords or API keys directly into your scripts. Instead, use environment variables or configuration files to store such information securely.

Script maintenance is an ongoing process. As systems, dependencies, and requirements change, your automation scripts may require updates. Regularly review and update your scripts to ensure they remain functional and efficient. Monitoring and logging can help you detect and address issues promptly.

Collaboration is often a part of script development, especially in team environments. Use collaboration tools like version control systems, code review platforms, and communication channels to work effectively with others. Code reviews can provide valuable feedback and improve the quality of your scripts.

Finally, don't forget about backups and disaster recovery plans. Automation scripts can have a significant impact on your workflow. Having backups of your scripts and a plan for recovering from unexpected failures can save you from potential data loss or downtime.

In summary, writing effective automation scripts is a combination of art and science. It involves understanding the problem, choosing the right tools, and following best practices in coding, testing, documentation, and security. Remember that automation scripts are not just utilities; they are valuable assets that can enhance your productivity and simplify complex tasks. With careful planning and craftsmanship, you can create automation scripts that make your life easier and more efficient. So, roll up your sleeves, embrace the scripting journey, and start automating your way to success!

Welcome to the world of scripting best practices and tips, where we'll explore ways to elevate your scripting skills and write code that's not only functional but also robust, maintainable, and a joy to work with. Scripting can be both an art and a science, and by following these best practices, you can make your scripts more effective and efficient.

First and foremost, let's talk about the importance of meaningful variable names. One of the golden rules of scripting is to choose descriptive variable names that convey the purpose of the variable. Instead of using vague names like "x" or "temp," opt for names that make your code self-explanatory. This practice not only helps you understand your code months or years down the line but also aids collaborators who might work with your script.

Comments are your allies in scripting. They provide context and explanations for your code. Whenever you write a piece of code that might not be immediately obvious to someone else (or even yourself in the future), add a comment to clarify its purpose or how it works. Good commenting habits make your scripts more accessible and maintainable.

Another essential practice is adhering to a consistent coding style. Whether you follow an established coding style guide (like PEP 8 for Python) or have your own, consistency matters. Consistent indentation, naming conventions, and code structure make your script easier to read and understand. If you're working in a team, adopting a shared coding style ensures that everyone is on the same page.

Keep your scripts modular by breaking them into smaller, reusable functions or modules. Modularization not only promotes code reuse but also simplifies testing and debugging. Each function should have a clear purpose and do one thing well. If a function becomes too long or complex, consider refactoring it into smaller functions.

Error handling is a critical aspect of scripting. Your script should gracefully handle errors and exceptions. Instead of letting errors crash your script, use try-except blocks (or equivalent error-handling mechanisms in your scripting language) to catch and handle exceptions. Think about how your script should behave when things go wrong, and implement appropriate error-handling strategies.

Automation scripts often involve working with files and external resources. When dealing with files, always close them after you're done. Leaving files open can lead to resource leaks and unexpected behavior. A common best practice is to use context managers (like **with** statements in Python) to ensure that files are properly closed, even if an exception occurs.

Documentation should not be an afterthought. It should be an integral part of your scripting process. Well-documented scripts are like a user manual, making it easier for others (and your future self) to understand and use your code. Explain the purpose of your script, how to run it, and any dependencies or prerequisites. Document your functions and their parameters, return values, and expected behavior.

Version control is your best friend in script development. Use a version control system like Git to track changes to your script over time. This not only provides a backup of your code but also enables collaboration with others. You can work on different branches, merge changes, and revert to previous versions if needed. Don't forget to commit your changes with meaningful commit messages.

Testing is a crucial part of scripting. Writing tests for your functions or scripts helps you catch bugs early and ensures that your code behaves as expected. Consider using testing frameworks or libraries relevant to your scripting language. Automate your tests to run them regularly, and add new tests when you make changes to your code. Testing provides confidence that your script is reliable.

Performance optimization can be important in scripting, especially for resource-intensive tasks. Profile your code to identify performance bottlenecks and areas for improvement. Techniques like caching, lazy loading, and algorithmic optimizations can help make your script more efficient. However, remember the rule of optimization: "Don't optimize prematurely." Focus on optimizing only when performance is a concern and you have evidence of a problem.

Security is a paramount concern when writing scripts, especially if they handle sensitive data or interact with external systems. Avoid hardcoding sensitive information like passwords or API keys directly into your scripts. Instead, use environment variables or configuration files to store such information securely. Be aware of common security vulnerabilities, like SQL injection or Cross-Site Scripting (XSS), and take steps to mitigate them.

When sharing your scripts with others, consider packaging and distribution. Depending on your scripting language, you might be able to create packages or modules that can be easily installed by others. Provide clear installation instructions and dependencies to make it as smooth as possible for users to get started with your script.

Backups and disaster recovery plans are not to be overlooked. Automation scripts can become critical components of your workflow. Regularly back up your scripts and any associated data. Consider what happens if your script fails unexpectedly or if you lose access to it. Having a plan for recovery can save you from potential data loss or downtime.

Lastly, embrace a growth mindset when it comes to scripting. The scripting landscape is constantly evolving, and new tools, libraries, and techniques emerge regularly. Stay curious and open to learning. Explore new scripting languages or frameworks that might be better suited to your tasks. Read and share knowledge with the scripting community.

In summary, scripting is both an art and a science, and following best practices can make your scripts more effective and maintainable. Choose meaningful variable names, use comments for clarity, adhere to a consistent coding style, and modularize your code. Handle errors gracefully, document your scripts, and use version control. Test your code, optimize for performance when necessary, and prioritize security. Consider packaging and distribution, and always have backup and recovery plans in place. Embrace continuous learning and growth in the world of scripting. With these practices in mind, you'll be well on your way to scripting success.

Chapter 9: Error Handling and Debugging

Welcome to the fascinating world of understanding Python errors and exceptions. As you journey through the realm of Python programming, encountering errors and exceptions is not a matter of if, but when. These little hiccups in your code are like signposts on the road to improvement, guiding you toward better, more reliable code.

So, what exactly are Python errors and exceptions? Well, think of them as messages from the Python interpreter, telling you that something unexpected has happened. These messages can range from simple reminders about missing parentheses to complex reports of issues deep within your code.

Errors in Python are broadly categorized into two main types: syntax errors and exceptions.

Syntax errors are the most basic type of error. They occur when you violate the rules of Python's syntax, such as forgetting a closing parenthesis, misspelling a variable name, or using an invalid operator. When you encounter a syntax error, Python will promptly inform you about the location of the problem and the nature of the error. These errors prevent your code from running at all, making them relatively easy to spot and fix.

Exceptions, on the other hand, are a bit more nuanced. Exceptions occur when your code is syntactically correct, but something unexpected happens during its execution. Python provides a variety of built-in exceptions, such as **TypeError**, **ValueError**, and **FileNotFoundError**, to help you pinpoint the issue. Exceptions can be triggered by a wide range of events, from trying to divide by zero to attempting to access a non-existent file.

Now, here's the good news: Python is designed to be forgiving and helpful. When an exception occurs, Python doesn't throw up its hands in defeat and crash your program. Instead, it raises an exception, which can be caught and handled by your code.

Let's say you're writing a program that prompts the user for their age, and the user decides to be mischievous and enters "chocolate" instead of a number. Python will raise a **ValueError** exception because it cannot convert "chocolate" to an integer. However, by anticipating this possibility and wrapping your code in a try-except block, you can gracefully handle the situation and prevent your program from crashing.

Here's an example of how you can use a try-except block to catch and handle exceptions:

pythonCopy code

```python
try: user_age = int(input("Please enter your age: ")) except ValueError: print("Invalid input. Please enter a valid age as a number.")
```

In this code, we attempt to convert the user's input to an integer. If the conversion fails and a **ValueError** exception is raised, our code catches the exception and prints an error message. This way, even if the user enters invalid input, your program remains responsive and user-friendly.

Python provides a wide range of built-in exceptions, each serving a specific purpose. For example, **IndexError** occurs when you try to access an index that doesn't exist in a list, while **FileNotFoundError** is raised when you attempt to open a file that doesn't exist.

You can also create your own custom exceptions by defining new exception classes. This is useful when you want to add a layer of specificity to your error-handling code. For instance, if you're building a program to manage a library, you might create a custom **BookNotFoundError** exception to handle cases where a requested book is not in the library's catalog.

Handling exceptions is a critical part of writing robust code. It allows your program to gracefully recover from unexpected situations and provide informative feedback to users. However, it's important to use exception handling judiciously. Catching too many exceptions or using overly broad catch-all blocks can make it challenging to debug issues in your code.

To handle exceptions effectively, it's essential to understand Python's exception hierarchy. At the top of the hierarchy is the **BaseException** class, which serves as the base class for all exceptions in Python. This means that every exception, including built-in ones like **ValueError** and **TypeError**, ultimately inherits from **BaseException**.

You can use this inheritance hierarchy to catch multiple exceptions in a single **except** block. For example, if you want to handle both **ValueError** and **TypeError** exceptions in the same way, you can catch them together:

pythonCopy code

```
try: user_input = input("Enter a number: ") result = 10 / int(user_input) except (ValueError, TypeError): print("Invalid input. Please enter a valid number.") except ZeroDivisionError: print("Cannot divide by zero.") except Exception as e: print(f"An unexpected error occurred: {e}")
```

In this example, we catch **ValueError** and **TypeError** exceptions together in the first **except** block. Then, we handle the **ZeroDivisionError** exception separately. Finally, we have a catch-all **except Exception** block that can catch any other exceptions not explicitly handled.

While catching multiple exceptions in a single **except** block can be convenient, it's crucial to be specific in your error handling. This helps you provide accurate feedback to the user and ensures that you're not masking unexpected issues by handling them too broadly.

Another useful feature in exception handling is the **finally** block. The **finally** block is executed regardless of whether an exception is raised or not. It's often used for cleanup tasks, such as closing files or releasing resources. For example:

```python
pythonCopy code
try: file = open("example.txt", "r") # Read data from the file except FileNotFoundError: print("File not found.") finally: file.close() # Ensure the file is always closed, even if an exception occurs.
```

In this code, the **finally** block ensures that the file is closed, even if an exception is raised while reading from it or if the file is not found.

In addition to handling exceptions with **try**, **except**, and **finally**, you can also raise exceptions explicitly using the **raise** statement. This can be useful when you want to indicate that something unexpected has occurred within your code. For example:

```python
pythonCopy code
def divide(a, b): if b == 0: raise ValueError("Cannot divide by zero.") return a / b try: result = divide(10, 0) except ValueError as e: print(f"An error occurred: {e}")
```

In this example, the **divide** function raises a **ValueError** if the denominator **b** is zero. When the exception is caught in the **try** block, we print an error message.

Understanding the specific exceptions that can be raised in Python and knowing when and how to handle them is a fundamental skill for any Python programmer. By effectively managing errors and exceptions, you can create more robust and reliable scripts and applications.

In summary, Python errors and exceptions are an integral part of programming. They are messages from the interpreter that help you identify and address issues in your code. Errors can be broadly categorized into syntax errors and exceptions, with exceptions being more nuanced and requiring careful handling. You can catch and handle exceptions using **try** and **except** blocks, and you can create your own custom exceptions for specialized error handling. It's important to be specific in your error handling and understand Python's exception hierarchy. The **finally** block is useful for cleanup tasks, and you can raise exceptions explicitly using the **raise** statement. Embrace errors and exceptions as opportunities to improve your code and make it more reliable. Happy coding!

Welcome to the world of debugging, a journey that every programmer embarks on to unravel the mysteries hidden within their code. Debugging is like detective work, and just like a seasoned detective, you'll need a set of tools and techniques to crack the case of the elusive bug.
Before we dive into the tools and techniques, let's demystify what debugging really is. At its core, debugging is the process of identifying and fixing errors or bugs in your code. These errors can manifest as unexpected behavior, crashes, or incorrect results. Debugging is an essential skill for every programmer because, as the saying goes, "To err is human."
So, where do you start when you suspect there's a bug lurking in your code? The first step is often the simplest: take a deep breath. Debugging can be frustrating, but staying calm and composed will help you approach the problem systematically. Remember, bugs are just puzzles waiting to be solved.
One of the most fundamental and effective debugging techniques is using print statements. When in doubt, print it

out! Inserting print statements in your code allows you to inspect the values of variables and track the flow of execution. This "print debugging" technique is like leaving breadcrumbs in the forest; it helps you retrace your code's steps and pinpoint where things go awry.

Here's an example of using print statements to debug a simple function that calculates the factorial of a number:

pythonCopy code

```
def factorial(n): result = 1 for i in range(1, n + 1): result *= i print(f"i: {i}, result: {result}") return result print(factorial(5))
```

By adding print statements inside the loop, you can see how the variables change with each iteration and identify any unexpected behavior.

But what if your code is more complex, and sprinkling it with print statements isn't practical? This is where integrated development environments (IDEs) come into play. Most modern IDEs offer robust debugging features that allow you to set breakpoints, inspect variables, and step through your code line by line.

Let's take a brief look at a typical debugging workflow in an IDE:

Setting Breakpoints: You can set breakpoints in your code at specific lines where you suspect the bug may be hiding. When your code reaches a breakpoint during execution, it pauses, allowing you to inspect the program's state.

Stepping Through Code: You can step through your code using commands like "step into," "step over," and "step out." These commands let you navigate through your code one line at a time, examining the values of variables as you go.

Inspecting Variables: Debuggers allow you to inspect the values of variables at runtime. You can view the current state of variables and even change their values to test different scenarios.

Call Stack: The call stack provides a trace of the function calls that led to the current point in your code. This is immensely helpful for understanding the flow of execution.

Watch Expressions: You can set watch expressions to monitor specific variables or expressions. The debugger will update these expressions' values as you step through your code.

Using an IDE's debugging features can significantly speed up the debugging process, especially in complex projects. It's like having a magnifying glass to examine every detail of your code.

In addition to print statements and IDEs, another powerful technique is writing test cases. Testing your code before and after making changes helps ensure that your fixes don't introduce new bugs. Unit tests, integration tests, and end-to-end tests are all valuable tools in your debugging arsenal. Testing frameworks like **unittest** in Python and **JUnit** in Java provide structured ways to create and run tests.

Let's say you're working on a function that calculates the square root of a number, and you suspect there might be a bug. You can write test cases that check the function's behavior against known inputs and expected outputs. If the tests fail after making changes, it's a clear signal that something went wrong.

Here's an example of a simple test case for a square root function using Python's **unittest** module:

pythonCopy code

```
import unittest def square_root(x): return x ** 0.5 class
TestSquareRoot(unittest.TestCase):                      def
test_positive_number(self):
self.assertEqual(square_root(9),        3.0)        def
```

test_negative_number(self): with self.assertRaises(ValueError): square_root(-4) if __name__ == '__main__': unittest.main()

In this example, we have two test cases: one for a positive number and another for a negative number. The tests use the **assertEqual** and **assertRaises** methods to check if the function behaves as expected.

Now, let's talk about some debugging tools that can come in handy. In addition to integrated debugging in IDEs, there are standalone debugging tools designed for specific programming languages.

For Python, one such tool is **pdb**, the Python Debugger. It's a built-in interactive debugger that you can invoke from the command line or within your code. **pdb** allows you to set breakpoints, step through code, and inspect variables interactively. Here's a simple example of using **pdb**:

pythonCopy code

```
import pdb def divide(a, b): result = a / b return result
pdb.set_trace() result = divide(10, 0)
```

When you run this code, it will pause execution at the **pdb.set_trace()** line and provide you with a command-line interface to inspect and debug your code.

In addition to **pdb**, there are third-party debugging tools and profilers available for Python, such as **PyCharm Debugger** and **Pyflame** for performance profiling.

For other programming languages like Java, you have debugging tools like **jdb** (Java Debugger) and various profilers like VisualVM.

Beyond tools and techniques, a crucial aspect of debugging is developing a systematic approach. Start by reproducing the bug consistently. Understand the conditions that trigger the bug, and create a minimal test case that exhibits the issue.

This simplified scenario makes it easier to isolate and fix the problem.

Next, take a scientific approach. Formulate hypotheses about what might be causing the bug, and systematically test each hypothesis. Don't be afraid to question your assumptions and be open to unexpected findings.

Version control systems are your allies in debugging. Use them to track changes in your code and, if needed, revert to a previous version when the bug wasn't present. This allows you to pinpoint when and where the bug was introduced.

Collaboration can be a powerful debugging technique. Discuss the problem with colleagues or seek help on online forums and communities. A fresh pair of eyes can often spot things you've missed.

Remember that debugging is a skill that improves with practice. Each bug you encounter is an opportunity to become a better debugger. Over time, you'll develop an intuition for where to look and what to test when issues arise.

In summary, debugging is an essential part of programming, and it's a skill that every developer should cultivate. Whether you're using print statements, IDEs, test cases, or specialized debugging tools, the goal is the same: to uncover and fix bugs in your code. Approach debugging systematically, formulate hypotheses, and use version control to your advantage. Collaborate with others and embrace each debugging challenge as a chance to level up your skills. Happy bug hunting!

Chapter 10: Practical Projects for Beginners

Welcome to the exciting world of building Python projects! You've come a long way in your journey of learning Python, from understanding the basics to mastering more advanced concepts. Now, it's time to put your skills to work and create real-world projects that showcase your abilities and solve practical problems.

Building your own Python projects is not only a great way to solidify your knowledge but also a rewarding experience. It allows you to apply what you've learned to create something tangible and useful. Plus, it's a fantastic way to demonstrate your programming skills to potential employers or clients.

So, where should you start when embarking on your first Python project? Let's walk through the process step by step.

Choose a Project Idea: The first step is to select a project idea that interests you. It could be something related to your hobbies, a problem you want to solve, or a tool that could make your daily life easier. The key is to choose something that motivates you, as you'll be spending time and effort on it.

Plan Your Project: Before you start coding, it's essential to plan your project. Define the project's goals, scope, and requirements. Break down the project into smaller tasks and create a timeline. Consider what technologies and libraries you'll need, as well as any potential challenges you might encounter.

Set Up Your Development Environment: Ensure you have a suitable development environment ready. You'll need a text editor or integrated development environment (IDE) to write your code. Popular choices for Python development include Visual Studio Code, PyCharm, and Jupyter Notebook.

Write Code Incrementally: Begin writing your code incrementally, one step at a time. Start with the most straightforward parts of your project and gradually build up to more complex features. This approach allows you to test and debug your code as you go, making it easier to catch and fix errors.

Use Version Control: Consider using a version control system like Git to track changes in your project. This helps you keep a history of your code, collaborate with others if needed, and revert to previous versions if something goes wrong.

Test Your Code: Testing is a critical part of project development. Write test cases to ensure that your code works as expected. Automated testing frameworks like **unittest** for Python can help you systematically test different parts of your project.

Debug and Refine: Expect to encounter bugs and unexpected behavior in your code. Debugging is the process of identifying and fixing these issues. Use debugging tools, print statements, and your problem-solving skills to troubleshoot and refine your code.

Documentation: Document your project as you go. Write clear comments in your code to explain its functionality. Create user documentation to help others understand how to use your project.

Optimization: Once your project works as intended, you can focus on optimization. Identify areas where your code can be made more efficient and improve its performance.

User Interface (If Applicable): If your project has a graphical user interface (GUI), design it to be user-friendly and intuitive. Tools like Tkinter for desktop apps or Flask/Django for web apps can help you create interfaces.

Deployment: Depending on your project's nature, you may need to deploy it to make it accessible to others. This could involve hosting a web application on a server, packaging a desktop application, or distributing a library.

Get Feedback: Share your project with friends, colleagues, or online communities to get feedback. Feedback can help you identify areas for improvement and make your project more robust.

Maintain and Update: Software projects are rarely "finished." Plan to maintain and update your project as needed. This may involve fixing bugs, adding new features, or adapting to changes in external libraries or platforms.

Now, let's explore some project ideas suitable for different skill levels:

Beginner Projects:

To-Do List App: Create a simple to-do list application that allows users to add, delete, and mark tasks as completed.

Number Guessing Game: Develop a game where the computer randomly selects a number, and the player tries to guess it.

Basic Calculator: Build a calculator that can perform basic arithmetic operations like addition, subtraction, multiplication, and division.

Intermediate Projects: 4. **Weather App:** Create a weather application that fetches weather data from an API and displays it to the user.

Personal Blog Website: Develop a personal blog website where you can write and publish articles.

Expense Tracker: Build an application to track income and expenses, providing insights into spending habits.

Advanced Projects: 7. **E-commerce Platform:** Create a full-fledged e-commerce platform with user authentication, product listings, shopping cart functionality, and payment processing.

Machine Learning Model Deployment: If you've dabbled in machine learning, deploy a machine learning model as a web service using tools like Flask or Django.

Game Development: Develop a 2D or 3D game using game development libraries like Pygame or Unity with Python scripting.

Remember that the complexity of your project should align with your skill level. It's perfectly fine to start with a beginner project and gradually work your way up to more advanced ones as you gain confidence and experience.

Throughout your journey of building Python projects, keep in mind that mistakes and challenges are part of the learning process. Don't be discouraged by errors or setbacks; instead, view them as opportunities to grow and improve your skills.

Building your own Python projects is not only educational but also immensely rewarding. As you create projects that solve real-world problems or bring joy to others, you'll discover the true power and versatility of Python as a programming language. So, roll up your sleeves, pick a project that excites you, and embark on your coding adventure. Your first Python project is just the beginning of an exciting journey in the world of software development. Happy coding!

Welcome to the world of hands-on projects, where you'll put your skills to the test and embark on exciting coding adventures. While learning the theory and fundamentals of programming is essential, it's through practical application that you truly solidify your knowledge and become a proficient developer.

Hands-on projects offer a unique opportunity to bridge the gap between theory and practice. They allow you to apply what you've learned in a real-world context, where you can see the tangible results of your efforts. These projects not only enhance your technical skills but also cultivate problem-solving abilities and boost your confidence as a programmer. Now, let's explore the benefits of hands-on projects and discover how they can accelerate your skill development.

Application of Concepts: Hands-on projects provide a platform to apply the programming concepts and techniques you've learned. Whether it's using loops and conditionals, working with databases, or building user interfaces, projects allow you to use these skills in practical scenarios.

Problem Solving: Programming is fundamentally about solving problems, and hands-on projects are problem-solving exercises in disguise. You'll encounter challenges and obstacles along the way, forcing you to think critically and devise creative solutions.

Coding Proficiency: Repetition is the key to mastering any skill, and coding is no exception. By consistently writing code and tackling projects, you reinforce your coding proficiency, making programming feel like second nature.

Project Diversity: The world of programming is vast and diverse, with applications in various domains. Hands-on projects offer the flexibility to choose projects that align with your interests, whether it's web development, data science, game development, or mobile app development.

Portfolio Building: As you complete hands-on projects, you build a portfolio of work that showcases your abilities to potential employers or clients. A well-documented portfolio can open doors to exciting opportunities.

Learning by Mistakes: Mistakes are an integral part of the learning process. When you encounter errors or bugs in your projects, you gain invaluable experience in debugging and troubleshooting—an essential skill for any programmer.

Creativity and Innovation: Hands-on projects encourage creativity and innovation. You have the freedom to explore new ideas, experiment with different technologies, and build projects that reflect your unique vision.

Now that you understand the importance of hands-on projects, let's delve into some project ideas that cater to different skill levels and areas of interest.

Beginner Projects:

Personal Portfolio Website: Create a personal website to showcase your skills, projects, and resume. You can use HTML, CSS, and a bit of JavaScript to make it interactive.

Task Tracker App: Build a task tracker application that allows users to add, update, and delete tasks. You can implement this as a web app using a Python web framework like Flask or Django.

Guess the Number Game: Develop a number guessing game where the computer selects a random number, and the player tries to guess it. This project is an excellent way to practice conditional statements and loops.

Intermediate Projects:

Weather Dashboard: Create a weather dashboard that fetches weather data from an API and displays it to users. This project can involve working with APIs, handling JSON data, and building a user-friendly interface.

E-commerce Website: Build a simple e-commerce website with features like product listings, a shopping cart, and user authentication. You can use web development frameworks like Ruby on Rails or Node.js with Express.js.

Chat Application: Create a real-time chat application using technologies like WebSocket or Socket.io. This project will give you experience with event-driven programming and building interactive, real-time systems.

Advanced Projects:

Machine Learning Model Deployment: If you've dabbled in machine learning, deploy a machine learning model as a web service. You can use Flask or Django to create a REST API that serves predictions.

Blockchain-Based App: Explore the world of blockchain by building a decentralized application (DApp). You can use blockchain platforms like Ethereum to create smart contracts and build decentralized applications.

Game Development: Develop a 2D or 3D game using game development engines like Unity or Unreal Engine. Game development is a challenging but rewarding field that combines programming with creativity.

As you embark on these projects, remember that the learning process is just as valuable as the final product. Here are some tips to make the most of your hands-on projects:

Plan and Research: Before diving into a project, take the time to plan and research the technologies and tools you'll need. Having a clear roadmap will save you time and frustration later.

Start Small: If you're new to a particular domain, start with smaller, manageable projects to build your confidence. As you gain experience, you can tackle more ambitious projects.

Use Version Control: Always use version control systems like Git to track changes in your code. It's a crucial skill that will serve you well in collaborative projects.

Document Your Work: Keep comprehensive documentation for your projects, including code comments, README files, and user guides. Clear documentation makes your work more accessible to others and helps you in the future.

Seek Feedback: Don't hesitate to share your projects with peers or online communities to get feedback. Constructive criticism can provide valuable insights and help you improve.

Stay Curious: The tech world is constantly evolving, so stay curious and open to learning new technologies and techniques. Hands-on projects are an excellent way to explore emerging trends.

In summary, hands-on projects are a powerful tool for skill development in the world of programming. They offer a practical way to apply your knowledge, solve real-world problems, and build a portfolio of work that reflects your abilities. Whether you're a beginner looking to build your first website or an experienced developer tackling complex challenges, there's a hands-on project waiting for you. So, roll up your sleeves, embrace the learning journey, and start coding your way to proficiency and innovation. Happy coding!

BOOK 2
PYTHON AUTOMATION MASTERY
INTERMEDIATE TECHNIQUES

ROB BOTWRIGHT

Chapter 1: Recap of Python Fundamentals

Welcome to a refresher on the fundamental concepts of Python, my friend. Whether you're a beginner just starting your Python journey or someone looking to reinforce their understanding, revisiting the basics is always a great idea. Python, known for its simplicity and readability, serves as an excellent foundation for building your programming skills.

Let's begin with the core of any programming language—variables. In Python, variables are like containers that hold data. You can store different types of data in variables, such as numbers, text, or even complex structures. To declare a variable, you simply choose a name and use the assignment operator (=) to assign a value to it.

pythonCopy code

greeting = "Hello, Python!" number = 42

Python, unlike some other languages, doesn't require you to explicitly declare the data type of a variable. It dynamically infers the data type based on the assigned value. This flexibility makes Python code concise and easy to write.

Once you have variables, you often need to perform operations on them. Python provides a set of operators for this purpose. For instance, you can use arithmetic operators like +, -, *, and / to perform basic math operations. You can also use the % operator for modulo (remainder) and ** for exponentiation.

pythonCopy code

result = 10 + 5 # Adds 10 and 5, result is 15 remainder = 17 % 4 # Computes the remainder when 17 is divided by 4, result is 1 power = 2 ** 3 # Raises 2 to the power of 3, result is 8

Python also supports comparison operators like == (equal to), != (not equal to), < (less than), > (greater than), <= (less than or equal to), and >= (greater than or equal to). These operators help you compare values and make decisions in your code.

pythonCopy code

```
is_equal = 5 == 5  # Checks if 5 is equal to 5, result is True
is_greater = 10 > 7  # Checks if 10 is greater than 7, result is
True  is_not_equal = "apple" != "orange"  # Checks if
"apple" is not equal to "orange", result is True
```

In Python, you can use conditional statements to control the flow of your code. The **if**, **elif** (short for "else if"), and **else** keywords help you create decision branches. Conditional statements allow your program to make choices and execute different code blocks based on certain conditions.

pythonCopy code

```
x = 10 if x > 5: print("x is greater than 5") elif x == 5:
print("x is equal to 5") else: print("x is less than 5")
```

Indentation is crucial in Python. Unlike some languages that use braces or parentheses to define blocks of code, Python uses indentation. Indentation helps maintain code readability and enforces a consistent coding style.

One of Python's strengths is its ability to work with collections of data. Lists, for example, are versatile and commonly used data structures. A list is an ordered collection of elements enclosed in square brackets. You can store various data types in a list, including numbers, strings, and even other lists.

pythonCopy code

```
fruits = ["apple", "banana", "cherry"] numbers = [1, 2, 3,
4, 5] mixed = [1, "apple", True, [5, 6, 7]]
```

To access elements in a list, you use indices. Python uses zero-based indexing, meaning the first element is at index 0, the second at index 1, and so on. You can also use negative indices to count from the end of the list.

pythonCopy code

```
fruits = ["apple", "banana", "cherry"] first_fruit = fruits[0]
# Accesses the first element, "apple" last_fruit = fruits[-1] #
Accesses the last element, "cherry"
```

Another powerful data structure is the dictionary. A dictionary is a collection of key-value pairs enclosed in curly braces. Each key is unique, and you can use it to retrieve the associated value quickly.

pythonCopy code

```
person = { "name": "Alice", "age": 30, "city": "New York"
} name = person["name"] # Retrieves the value associated
with the key "name", which is "Alice"
```

Python also offers loops to iterate through collections and perform repetitive tasks. The **for** loop is used to iterate over elements in a sequence, such as a list or a string.

pythonCopy code

```
fruits = ["apple", "banana", "cherry"] for fruit in fruits:
print(fruit)
```

The **while** loop, on the other hand, is used for repetitive tasks as long as a certain condition is met.

pythonCopy code

```
count = 0 while count < 5: print(count) count += 1
```

Functions are reusable blocks of code that can take inputs, perform tasks, and return outputs. Python provides built-in functions like **print()** and **len()**, but you can also define your own functions.

pythonCopy code

```python
def greet(name): return f"Hello, {name}!" message = greet("Alice") # Calls the greet function and assigns the result to the message variable
```

Functions are a fundamental concept in Python and play a crucial role in structuring your code. They allow you to break your program into smaller, more manageable parts.

Exception handling is another vital aspect of Python programming. Errors can occur in your code, and you can use **try, except**, and **finally** blocks to handle them gracefully.
pythonCopy code

```python
try: result = 10 / 0 # Attempting to divide by zero except ZeroDivisionError as e: print(f"Error: {e}") finally: print("This block always executes")
```

In Python, you have access to a vast standard library that provides ready-made modules and functions to perform a wide range of tasks. Whether it's working with files, handling dates and times, or performing network operations, the Python standard library has you covered.

Now that we've refreshed your memory on these fundamental Python concepts, you're well-prepared to tackle more advanced topics and dive into exciting projects. Remember, practice is key to becoming a proficient programmer. Keep coding, exploring, and building, my friend. The Python world is full of possibilities, waiting for you to explore them.

Welcome to the next phase of your Python journey, where we'll delve into essential concepts for intermediate Python programming. By this point, you've likely mastered the basics of Python and are ready to explore more advanced topics that will empower you to tackle complex projects and enhance your programming skills.

One crucial concept to grasp as an intermediate Python programmer is **object-oriented programming (OOP)**. Python is an object-oriented language, which means it's built around the concept of objects and classes. Objects are instances of classes, and classes are like blueprints that define the structure and behavior of objects.

Let's break it down a bit further. Imagine you're building a program to model a zoo. You might have different types of animals, such as lions, tigers, and bears. In an object-oriented approach, you'd create a class for each type of animal with attributes (e.g., name, age) and methods (e.g., eat, sleep) that define their behavior. Then, you can create instances (objects) of those classes for each animal in the zoo.

pythonCopy code

```
class Animal: def __init__(self, name, age): self.name =
name self.age = age def eat(self): print(f"{self.name} is
eating.") class Lion(Animal): def roar(self):
print("Roar!") # Create instances of the Lion class simba =
Lion("Simba", 5) nala = Lion("Nala", 4) simba.eat() #
Simba is eating. nala.roar() # Roar!
```

In this example, we have an **Animal** class that serves as the base class for all animals. The **Lion** class inherits from **Animal** and adds its own unique method, **roar**. When we create instances of the **Lion** class, they inherit the attributes and methods from **Animal** and can also access their specialized **roar** method.

OOP provides a structured way to organize and model real-world systems, making your code more modular, reusable, and easier to understand. It's a powerful paradigm that's widely used in Python development.

Another intermediate concept is **working with external data**. In many real-world applications, you'll need to interact with external data sources, such as databases, files, or web APIs. Python offers various libraries and modules to make this process smooth and efficient.

For instance, to work with databases, you can use the popular library called **SQLAlchemy**. SQLAlchemy allows you to interact with relational databases like MySQL, PostgreSQL, and SQLite using Python code. You can define database tables as Python classes, query the database using Python methods, and manage data seamlessly.

pythonCopy code

```
from sqlalchemy import create_engine, Column, Integer, String from sqlalchemy.orm import sessionmaker from sqlalchemy.ext.declarative import declarative_base # Define a SQLite database and create a session engine = create_engine("sqlite:///mydatabase.db") Session = sessionmaker(bind=engine) session = Session() # Define a Python class that corresponds to a database table Base = declarative_base() class User(Base): __tablename__ = "users" id = Column(Integer, primary_key=True) name = Column(String) age = Column(Integer) # Create the table in the database Base.metadata.create_all(engine) # Add a new user to the database new_user = User(name="Alice", age=30) session.add(new_user) session.commit() # Query the database user = session.query(User).filter_by(name="Alice").first()
print(user.name) # Output: Alice
```

Working with files is another crucial skill. Python offers built-in functions for file operations. You can read from and write to text files, binary files, and even work with CSV or JSON data formats. These capabilities are essential for tasks like data analysis, log processing, or configuration management. pythonCopy code

```
# Reading from a text file  with open("myfile.txt", "r") as file: content = file.read() print(content) # Writing to a text file  with  open("myoutput.txt",  "w")  as  file: file.write("Hello, Python!") # Working with CSV data  import csv with open("data.csv", "r") as csv_file: csv_reader = csv.reader(csv_file) for row in csv_reader: print(row) # Working with JSON data  import json data = {"name": "Alice",  "age":  30} json_string  =  json.dumps(data) print(json_string) # Output: {"name": "Alice", "age": 30}
```

Python's **modules and packages** are another critical aspect of intermediate programming. Modules are files containing Python code that you can import into your programs. Packages are collections of related modules organized in directories. They allow you to structure your code and manage complexity effectively.

You can create your own modules and packages to encapsulate functionality and promote code reusability. Moreover, Python's extensive standard library provides a wide range of modules for various purposes, from working with dates and times (using **datetime**) to performing regular expressions (with **re**) or handling command-line arguments (using **argparse**).

When building larger and more complex applications, **testing** becomes indispensable. Testing ensures that your code works as expected and helps you catch and fix issues early in the development process. Python offers several testing frameworks, including **unittest**, **pytest**, and **nose**. These frameworks allow you to write test cases to validate the correctness of your code.

```python
pythonCopy code
import unittest def add(a, b): return a + b class
TestAddition(unittest.TestCase):                    def
test_add_positive_numbers(self): result = add(3, 5)
self.assertEqual(result,              8)              def
test_add_negative_numbers(self): result = add(-2, -4)
self.assertEqual(result, -6) if __name__ == "__main__":
unittest.main()
```

In the example above, we define a simple function **add** and create test cases using **unittest**. Each test case method begins with "test_", and we use assertion methods like **assertEqual** to check if the function's output matches the expected result.

Lastly, as you advance in Python, you'll often encounter the need for **concurrency and parallelism**. Concurrency allows you to manage multiple tasks that run independently but potentially share resources, while parallelism involves executing multiple tasks simultaneously to improve performance. Python provides libraries like **threading** and **multiprocessing** for managing concurrency and parallelism.

The **threading** module allows you to work with threads, which are lightweight, independently executing units of a process. They are suitable for tasks that involve I/O operations, like reading and writing to files or making network requests.

```python
pythonCopy code
import threading def print_numbers(): for i in range(1,
6): print(f"Number: {i}") def print_letters(): for letter in
"ABCDE": print(f"Letter: {letter}") # Create two threads
thread1 = threading.Thread(target=print_numbers) thread2
= threading.Thread(target=print_letters) # Start the threads
```

thread1.start() thread2.start() # Wait for the threads to finish thread1.join() thread2.join() print("Both threads have finished.")

The **multiprocessing** module, on the other hand, allows you to work with multiple processes, which are independent and have their own memory space. This makes them suitable for CPU-bound tasks that benefit from true parallelism.

pythonCopy code

import multiprocessing def square(number): result = number * number print(f"The square of {number} is {result}") if __name__ == "__main__": numbers = [1, 2, 3, 4, 5] pool = multiprocessing.Pool(processes=2) pool.map(square, numbers) pool.close() pool.join()

In this example, we use the **multiprocessing** module to square a list of numbers concurrently, leveraging the power of multiple processes.

These intermediate concepts lay the groundwork for more advanced Python programming. As you continue your Python journey, you'll encounter topics such as web development with frameworks like Django and Flask, data analysis and machine learning with libraries like NumPy and TensorFlow, and much more. Embrace the learning process, practice regularly, and don't hesitate to explore new challenges and projects. Python's versatility and robust community support make it an exciting language to master. Happy coding!

Chapter 2: Advanced Data Structures

Welcome to the fascinating world of Python's data structures! In this chapter, we'll embark on a journey to explore the essential building blocks that empower you to manipulate and organize data effectively. These data structures serve as the backbone of countless software applications, from simple scripts to complex systems. As an intermediate Python programmer, understanding them deeply will significantly enhance your capabilities.

Let's start our exploration with the versatile and fundamental **lists**. Think of a list as a container that can hold a sequence of elements, allowing you to store and access data in a structured manner. Lists are incredibly flexible—you can mix different data types within a single list, making them a go-to choice for various scenarios.

pythonCopy code

```
fruits = ["apple", "banana", "cherry"] numbers = [1, 2, 3, 4, 5] mixed = [1, "apple", True, [5, 6, 7]]
```

Accessing elements in a list is straightforward. Python uses zero-based indexing, which means the first element is at index 0, the second at index 1, and so on. You can also use negative indices to count from the end of the list.

pythonCopy code

```
fruits = ["apple", "banana", "cherry"] first_fruit = fruits[0] # Accesses the first element, "apple" last_fruit = fruits[-1] # Accesses the last element, "cherry"
```

Lists provide various methods to manipulate their contents. You can add elements using **append()** or **insert()**, remove elements using **remove()** or **pop()**, and even sort the list with **sort()**.

```python
Copy code
fruits = ["apple", "banana", "cherry"]
fruits.append("orange") # Adds "orange" to the end of the
list fruits.insert(1, "grape") # Inserts "grape" at index 1
fruits.remove("banana") # Removes "banana" from the list
popped_fruit = fruits.pop(2) # Removes and returns the
element at index 2 fruits.sort() # Sorts the list in
alphabetical order
```

Next, we have **tuples**, which are similar to lists but with a significant difference: they are immutable. Once you create a tuple, you cannot change its contents. This immutability makes tuples suitable for situations where you need to ensure that the data remains constant.

```python
Copy code
dimensions = (10, 20, 30) # Creating a tuple x, y, z =
dimensions # Unpacking a tuple into variables
```

Tuples are commonly used to represent collections of related values. For example, you might use a tuple to store the latitude and longitude of a location or the RGB color values of a pixel.

Now, let's shift our attention to **dictionaries**. Dictionaries are collections of key-value pairs. Each key in a dictionary is unique and maps to a specific value. Think of a dictionary as a real-world dictionary where words (keys) are associated with their meanings (values).

```python
Copy code
person = { "name": "Alice", "age": 30, "city": "New York"
} name = person["name"] # Retrieves the value associated
with the key "name", which is "Alice"
```

Dictionaries are efficient for quick data retrieval, as you can access values by their keys in constant time. They are handy for tasks like storing configuration settings, building data caches, or organizing data in a structured way.

In Python, we also have **sets**, which are collections of unique elements. Sets are useful when you want to work with a collection of items while ensuring that each item appears only once.

pythonCopy code

```
fruits = {"apple", "banana", "cherry"} fruits.add("orange")
# Adds "orange" to the set  fruits.remove("banana")  #
Removes "banana" from the set
```

Sets are particularly valuable when you need to perform operations like union, intersection, or difference between collections. They provide efficient methods for these operations.

Another essential data structure is the **string**. While strings may seem simple, they are incredibly powerful and versatile. In Python, strings are sequences of characters enclosed in single, double, or triple quotes.

pythonCopy code

```
text = "Hello, Python!"
```

Strings support numerous operations like concatenation, slicing, and searching. You can manipulate and analyze text data with ease using these operations.

pythonCopy code

```
greeting = "Hello" name = "Alice" message = greeting + ", "
+ name + "!"  # Concatenating strings  sentence = "This is a
sample sentence."  substring = sentence[10:16]  # Slicing
the  string  to  extract  "sample"  position  =
sentence.find("sample") # Finding the position of "sample"
in the string
```

Beyond these fundamental data structures, Python also offers more specialized collections, such as **queues, stacks,** and **heaps**, through the **collections** module. These structures are essential for solving specific problems efficiently.

For example, a **queue** is a collection that follows the First-In-First-Out (FIFO) principle. You can use it for tasks like managing tasks in a job queue or ensuring the order of processing in a data pipeline.

pythonCopy code

```
from collections import deque queue = deque()
queue.append("task1") queue.append("task2") next_task =
queue.popleft() # Removes and returns the first task in the
queue
```

On the other hand, a **stack** follows the Last-In-First-Out (LIFO) principle. You might use a stack for tasks like evaluating expressions or tracking the state of a game.

pythonCopy code

```
stack = [] stack.append("item1") stack.append("item2")
last_item = stack.pop() # Removes and returns the last item
in the stack
```

Heaps are specialized trees that allow efficient retrieval of the smallest (or largest) element. They are useful for tasks like priority queues or finding the k smallest elements in a dataset.

pythonCopy code

```
import heapq numbers = [3, 1, 4, 1, 5, 9, 2, 6, 5, 3]
heapq.heapify(numbers) # Converts the list into a heap
smallest = heapq.heappop(numbers) # Removes and returns
the smallest element
```

In Python, you can also define your own custom data structures using classes. By creating classes, you can encapsulate data and behavior into objects, making your code more organized and easier to maintain.

pythonCopy code

```python
class Point: def __init__(self, x, y): self.x = x self.y = y def
distance_to_origin(self): return (self.x ** 2 + self.y ** 2)
** 0.5 point = Point(3, 4) distance =
point.distance_to_origin() # Calculates the distance to the
origin
```

As you continue to explore and experiment with these data structures, you'll gain a deeper understanding of their strengths and use cases. Each structure has its unique characteristics and is suited to particular problems. Mastering them equips you with the tools to design elegant and efficient solutions to a wide range of challenges in the world of programming.

In our ongoing journey through Python's intermediate concepts, we'll delve into more advanced topics, including object-oriented programming principles, data manipulation techniques, and advanced algorithms. The world of Python programming is vast, and each concept you master brings you closer to becoming a proficient Python programmer. So, keep exploring, keep coding, and enjoy the journey ahead!

Welcome to a new dimension of Python programming where you have the power to create custom data structures tailored to your specific needs. In this chapter, we'll explore the fascinating world of custom data structures, understand how they work, and discover the incredible flexibility they offer for solving complex problems.

Python, known for its simplicity and versatility, allows you to create custom data structures by defining classes. These classes serve as blueprints for your data structures, specifying the attributes and methods that define their behavior. When you create an instance of a class, you're essentially creating an object that embodies the characteristics and capabilities defined by that class.

Let's begin our exploration by creating a custom data structure known as a **linked list**. A linked list is a collection of nodes, where each node contains data and a reference (or link) to the next node in the sequence. Linked lists are versatile and efficient for dynamic data storage.

pythonCopy code

```
class Node: def __init__(self, data): self.data = data
self.next = None class LinkedList: def __init__(self):
self.head = None def append(self, data): new_node =
Node(data) if not self.head: self.head = new_node return
current = self.head while current.next: current =
current.next current.next = new_node
```

In this example, we define two classes: **Node** to represent individual elements in the linked list and **LinkedList** to manage the list as a whole. The **LinkedList** class has a method **append** that adds new nodes to the end of the list.

Creating a linked list allows you to dynamically manage data in a way that conventional arrays or lists might not. Linked lists are particularly useful when you need to insert or delete elements frequently, as you can do so with minimal overhead.

Another custom data structure that comes in handy is the **stack**. A stack is a Last-In-First-Out (LIFO) data structure, meaning that the most recently added item is the first to be removed. You might have encountered stacks when dealing with function calls or tracking states in algorithms.

pythonCopy code

```
class Stack: def __init__(self): self.items = [] def
push(self, item): self.items.append(item) def pop(self): if
not self.is_empty(): return self.items.pop() def
is_empty(self): return len(self.items) == 0
```

The **Stack** class has methods for pushing items onto the stack (**push**), popping items from the stack (**pop**), and checking if the stack is empty (**is_empty**). Stacks are often used for managing program flow, undo functionality, or parsing expressions.

On the flip side of the coin, we have the **queue**, which follows the First-In-First-Out (FIFO) principle. Queues are crucial for scenarios where you need to manage tasks in a sequential manner, like processing requests in a web server or managing print jobs in a printer queue.

pythonCopy code

```python
class Queue: def __init__(self): self.items = [] def enqueue(self, item): self.items.insert(0, item) def dequeue(self): if not self.is_empty(): return self.items.pop() def is_empty(self): return len(self.items) == 0
```

The **Queue** class allows you to enqueue items (**enqueue**) and dequeue items (**dequeue**) while ensuring that the first item added is the first to be removed. This behavior makes queues invaluable for maintaining order in various applications.

Moving on, let's dive into the world of **trees**. Trees are hierarchical data structures consisting of nodes connected by edges. Each tree has a root node, which serves as the starting point, and nodes are organized into levels, with child nodes branching out from parent nodes.

You can create custom tree structures for a multitude of purposes. One common type is the **binary search tree (BST)**. In a BST, each node has at most two children: a left child and a right child. Nodes are arranged in a way that allows for efficient searching, insertion, and deletion of elements.

pythonCopy code

```python
class TreeNode: def __init__(self, key): self.key = key
self.left = None self.right = None class BinarySearchTree:
def __init__(self): self.root = None def insert(self, key):
self.root = self._insert_recursive(self.root, key) def
_insert_recursive(self, root, key): if not root: return
TreeNode(key) if key < root.key: root.left =
self._insert_recursive(root.left, key) else: root.right =
self._insert_recursive(root.right, key) return root
```

In this example, we define two classes: **TreeNode** for the nodes in the BST and **BinarySearchTree** to manage the tree structure. The **insert** method adds new nodes while maintaining the binary search tree properties.

Custom trees like BSTs have numerous applications, from organizing data for efficient retrieval to implementing data structures like sets and maps. They are also used extensively in algorithms for tasks like sorting and searching.

Custom data structures become even more powerful when combined with algorithms. For instance, you can implement **graph** data structures and algorithms to solve problems in areas such as social network analysis, route planning, and recommendation systems.

Custom graphs can represent complex relationships between entities, and algorithms like breadth-first search (BFS) and depth-first search (DFS) can help you traverse and analyze these relationships effectively.

pythonCopy code

```python
class Graph: def __init__(self): self.vertices = {} def
add_vertex(self, vertex): self.vertices[vertex] = [] def
add_edge(self, from_vertex, to_vertex): if from_vertex in
self.vertices: self.vertices[from_vertex].append(to_vertex)
```

In this snippet, we define a **Graph** class for creating and managing graph structures. The **add_vertex** method adds vertices, and the **add_edge** method adds edges between vertices.

As you can see, custom data structures open up a world of possibilities for solving complex problems. Whether you're modeling real-world scenarios, optimizing algorithms, or designing efficient data storage, the ability to create tailored data structures is a valuable skill in your programming arsenal.

In our ongoing journey through intermediate Python concepts, we'll continue to explore advanced topics, including algorithms, object-oriented programming principles, and real-world applications. The world of Python programming is rich and diverse, offering endless opportunities to innovate and create. So, keep exploring, keep coding, and embrace the art of designing custom data structures to conquer the challenges that lie ahead!

Chapter 3: Object-Oriented Programming with Python

Welcome to the world of object-oriented programming (OOP), a powerful paradigm that empowers you to model and design software in a more organized and intuitive way. In this chapter, we'll delve into the fundamental concepts of OOP and explore how they can enhance your Python programming skills.

At the heart of OOP lies the concept of an **object**, which is a self-contained unit that combines data (attributes) and behaviors (methods). Objects are the building blocks of OOP, and they allow you to represent real-world entities in your code. For example, in a car rental system, you can create objects to represent cars, customers, and rental transactions.

Classes, on the other hand, are blueprints or templates for creating objects. Think of a class as a cookie cutter, and objects as the cookies you create using that cutter. A class defines the attributes and methods that its objects will have.

In Python, you create a class using the **class** keyword, followed by the class name and a colon. Inside the class, you define its attributes and methods.

pythonCopy code

```python
class Car: def __init__(self, make, model, year): self.make = make self.model = model self.year = year def start_engine(self): print(f"The {self.make} {self.model}'s engine is running.")
```

In this example, we define a **Car** class with attributes like **make**, **model**, and **year**. The **start_engine** method simulates starting the car's engine.

When you create an instance of a class, you're creating an object. This process is called **instantiation**. You can create multiple objects (car instances) from the same class, each with its own unique data.

pythonCopy code

```
my_car = Car("Toyota", "Camry", 2022) your_car = Car("Ford", "Mustang", 2023)
```

In this snippet, we've created two car objects, **my_car** and **your_car**, each with its own set of attributes.

One of the key concepts in OOP is **encapsulation**, which refers to the bundling of data and methods that operate on that data into a single unit, i.e., the class. Encapsulation allows you to control access to an object's data, ensuring that it's modified or accessed only through well-defined methods.

In our **Car** class, we encapsulate the car's attributes and behavior. The **start_engine** method encapsulates the process of starting the engine, hiding the internal details from the outside world.

Inheritance is another vital concept in OOP. It enables you to create a new class (a **subclass** or **derived class**) based on an existing class (a **base class** or **parent class**). Subclasses inherit attributes and methods from their parent class, and they can also have their own unique attributes and methods. Consider a scenario where you want to model different types of cars, like electric cars. You can create a subclass **ElectricCar** that inherits from the **Car** class.

pythonCopy code

```
class ElectricCar(Car): def __init__(self, make, model, year, battery_capacity): super().__init__(make, model, year) self.battery_capacity = battery_capacity def charge_battery(self): print(f"Charging the {self.make} {self.model}'s battery.")
```

In this example, **ElectricCar** inherits the **make, model,** and **year** attributes from the **Car** class and introduces a new attribute, **battery_capacity**. It also has its own method, **charge_battery**.

Inheritance promotes code reusability, allowing you to build on existing classes and create more specialized classes without duplicating code. It reflects the real-world "is-a" relationship, where an electric car "is-a" car.

A powerful concept that complements inheritance is **polymorphism**, which means "many forms." Polymorphism allows objects of different classes to be treated as objects of a common base class. It enables you to write code that can work with objects from various classes, as long as they share a common interface (attributes and methods).

For example, both **Car** and **ElectricCar** have a **start_engine** method. By using polymorphism, you can call this method on objects of either class without knowing their specific types.

```python
pythonCopy code
def       start_car_engine(car):        car.start_engine()
start_car_engine(my_car)  # Calls start_engine on a Car
object start_car_engine(your_car) # Calls start_engine on an
ElectricCar object
```

In this code, the **start_car_engine** function accepts any object with a **start_engine** method, regardless of whether it's a **Car** or an **ElectricCar**. This is the essence of polymorphism.

Python also supports another form of polymorphism called **method overriding**, where a subclass provides its own implementation of a method defined in the parent class. When you call the overridden method on a subclass object, the subclass's implementation is executed.

In our **ElectricCar** class, we can override the **start_engine** method to add behavior specific to electric cars.

```python
pythonCopy code
class ElectricCar(Car): def __init__(self, make, model,
year, battery_capacity): super().__init__(make, model,
year) self.battery_capacity = battery_capacity def
start_engine(self): print(f"The {self.make} {self.model}'s
electric motor is running.") def charge_battery(self):
print(f"Charging the {self.make} {self.model}'s battery.")
```

Now, when you call **start_engine** on an **ElectricCar** object, it
uses the overridden method defined in the **ElectricCar** class,
providing a customized message.

In the world of OOP, you'll often encounter the term
composition, which is an alternative to inheritance.
Composition allows you to build complex objects by
combining simpler objects as components. This approach
promotes code reusability and flexibility while avoiding some
of the pitfalls of deep inheritance hierarchies.

As you dive deeper into OOP, you'll also encounter concepts
like **abstraction** (hiding complex implementation details),
interfaces (specifying a contract for classes), and **design
patterns** (reusable solutions to common programming
problems).

Understanding and applying these concepts will elevate your
Python programming skills and empower you to design
elegant, maintainable, and scalable software. Embrace OOP
as a powerful tool in your programming toolbox, and let it
guide you in building robust and efficient applications.
Happy coding! Now that you've gained a solid understanding
of the fundamental concepts of object-oriented
programming (OOP), it's time to explore how you can
implement these principles effectively in Python. In this
chapter, we'll dive deeper into the practical aspects of OOP
and see how Python's syntax and features make it a fantastic
language for OOP.

Let's start by revisiting the four key pillars of OOP: encapsulation, inheritance, polymorphism, and abstraction. These principles serve as the foundation for designing organized and maintainable code.

Encapsulation is all about bundling data (attributes) and methods that operate on that data within a class. In Python, you can control access to an object's attributes by defining them as private or protected using the underscore _ convention. Private attributes are indicated by a double underscore __.

For example, you can modify our **Car** class to have a private attribute called __**mileage.**

pythonCopy code

```
class Car: def __init__(self, make, model, year): self.make = make self.model = model self.year = year self.__mileage = 0 def start_engine(self): print(f"The {self.make} {self.model}'s engine is running.") def update_mileage(self, mileage): if mileage >= 0: self.__mileage = mileage else: print("Mileage cannot be negative.") def get_mileage(self): return self.__mileage
```

Here, __**mileage** is a private attribute that can only be accessed or modified through methods like **update_mileage** and **get_mileage**. This encapsulation ensures data integrity and allows you to control how the attribute is modified.

Inheritance, as we discussed earlier, allows you to create new classes based on existing ones. In Python, you can inherit from a base class by specifying the parent class in parentheses when defining the child class.

pythonCopy code

```
class ElectricCar(Car): def __init__(self, make, model, year, battery_capacity): super().__init__(make, model, year) self.__battery_capacity = battery_capacity def
```

charge_battery(self): print(f"Charging the {self.make} {self.model}'s battery.") def update_mileage(self, mileage): print("Electric cars do not have mileage.")

In this example, **ElectricCar** is a subclass of **Car**. It inherits the attributes and methods of **Car** while also introducing its own attributes like **__battery_capacity** and methods like **charge_battery**. Additionally, it overrides the **update_mileage** method to provide a specialized implementation.

Polymorphism is a powerful concept that allows objects of different classes to be treated as objects of a common base class. Python makes use of polymorphism through dynamic typing and duck typing.

Dynamic typing means that you don't need to declare the type of a variable; Python determines it at runtime. This flexibility allows you to use the same function or method with different types of objects.

For instance, consider a **drive** function that takes a car object and starts its engine, regardless of whether it's a **Car** or an **ElectricCar**.

pythonCopy code

```
def drive(car): car.start_engine() drive(my_car)  # Calls
start_engine on a Car object  drive(your_car)  # Calls
start_engine on an ElectricCar object
```

Both **my_car** and **your_car** can be passed to the **drive** function because they share a common method **start_engine**. This is polymorphism in action.

Duck typing, a concept closely related to dynamic typing, emphasizes the importance of an object's behavior over its class or type. If an object walks like a duck and quacks like a duck, it's treated as a duck. In Python, this means you can use objects based on what they can do, rather than what they are.

Abstraction involves hiding complex implementation details and providing a simplified interface for interacting with objects. In Python, you can create abstract classes and methods using the **abc** (Abstract Base Classes) module.

pythonCopy code

```
from abc import ABC, abstractmethod class Vehicle(ABC):
@abstractmethod def start_engine(self): pass class
Car(Vehicle): def start_engine(self): print("Car engine
started.") class ElectricCar(Vehicle): def
start_engine(self): print("Electric car motor running.")
```

In this example, **Vehicle** is an abstract class with an abstract method **start_engine**. Subclasses like **Car** and **ElectricCar** must implement this method. Abstraction allows you to define a common interface while leaving the specifics to the implementing classes.

In Python, you can also use **interfaces** as a form of abstraction. An interface defines a contract that classes must adhere to, specifying a set of methods that must be implemented. While Python doesn't have a built-in keyword for interfaces, you can achieve this using abstract base classes or simply by documenting the expected methods in the class docstring.

Now, let's talk about **method resolution order** (MRO), a crucial concept when dealing with inheritance in Python. MRO defines the order in which Python searches for methods in a class hierarchy.

Python uses the C3 linearization algorithm to determine the MRO. It ensures that methods are resolved in a consistent and predictable way, even in complex multiple inheritance scenarios.

pythonCopy code

```
class A: def show(self): print("A") class B(A): def
show(self): print("B") class C(A): def show(self):
print("C") class D(B, C): pass d = D() d.show() # Output: B
```

In this example, the **D** class inherits from both **B** and **C**, which in turn inherit from **A**. The MRO ensures that the **show** method of class **B** takes precedence when **d.show()** is called. Lastly, Python supports a feature called **decorators** that can be used for managing access control to methods and attributes. Decorators are functions that modify the behavior of other functions or methods. They are often used to add functionality like logging, authentication, or access control to methods.

pythonCopy code

```
def admin_only(func): def wrapper(self, *args,
**kwargs): if self.is_admin: return func(self, *args,
**kwargs) else: raise PermissionError("Only admin can
access this method.") return wrapper class Employee: def
__init__(self, name, is_admin=False): self.name = name
self.is_admin = is_admin @admin_only def
delete_employee(self): print(f"{self.name} has been
deleted.") employee1 = Employee("Alice", is_admin=True)
employee2 = Employee("Bob")
employee1.delete_employee() # Successfully deletes Alice
employee2.delete_employee() # Raises PermissionError for
Bob
```

In this example, the **admin_only** decorator restricts access to the **delete_employee** method based on the **is_admin** attribute. If the employee is an admin (**is_admin=True**), they can delete other employees; otherwise, permission is denied.

Chapter 4: Working with Libraries and Packages

Welcome to the exciting world of leveraging external libraries and packages in your Python projects. In this chapter, we'll explore how you can harness the power of third-party libraries to enhance your Python applications and save valuable development time.

As you embark on your Python programming journey, you'll soon discover that Python's strength lies not only in its simplicity and readability but also in its vast ecosystem of libraries and packages. These external resources, created by talented developers worldwide, provide ready-made solutions to a wide range of problems and challenges.

Let's start by understanding what libraries and packages are and how they can benefit your Python projects. In the context of programming, a **library** is a collection of pre-written code or modules that you can use to perform specific tasks or functions. A **package** is a more extensive collection of libraries, often organized into a directory structure, that serves a particular purpose or domain.

Python's Package Index (PyPI) is a treasure trove of libraries and packages, covering areas like web development, data analysis, machine learning, game development, and much more. Whether you're building a web application, conducting data analysis, or working on an artificial intelligence project, chances are there's a library or package that can streamline your work.

To harness the power of these external resources, you first need to understand how to **install** them. Python's package manager, **pip**, is your go-to tool for this task. With a simple command like **pip install package_name**, you can download and install the desired library or package effortlessly.

For example, if you're working on a web development project using the Flask framework, you can install Flask with a single command:

pythonCopy code

pip install Flask

Once the library or package is installed, you can **import** it into your Python code using the **import** statement. This allows you to access the functions, classes, and modules provided by the library within your project.

pythonCopy code

```
import flask app = flask.Flask(__name__) @app.route('/')
def hello_world(): return 'Hello, World!' if __name__ ==
'__main__': app.run()
```

In this example, we've imported Flask and used it to create a simple web application that responds with "Hello, World!" when accessed in a web browser.

Now, let's explore some of the key benefits of leveraging external libraries and packages in your Python projects:

Increased Productivity: External libraries and packages save you time and effort by providing pre-built solutions for common tasks. Instead of reinventing the wheel, you can focus on solving the unique challenges of your project.

Code Quality: Many popular libraries are well-maintained and thoroughly tested by a community of developers. By using these libraries, you can benefit from their reliability and robustness, ensuring the quality of your code.

Scalability: Libraries and packages are designed to scale with your project. Whether you're building a small script or a large-scale application, these resources can adapt to your needs.

Community Support: The Python community is vibrant and welcoming. When you encounter issues or have questions about a library or package, you can often find answers in forums, documentation, or community discussions.

Specialized Functionality: Some libraries are highly specialized, offering advanced features in specific domains. For instance, libraries like NumPy and pandas excel in data manipulation and analysis, while TensorFlow and PyTorch are renowned for deep learning and machine learning tasks.

Interoperability: Python is known for its versatility and ability to integrate with other programming languages and systems. Many libraries and packages facilitate seamless integration with technologies like C/C++, Java, and more.

Open Source: The majority of Python libraries and packages are open source, meaning you can inspect their source code, contribute to their development, and use them freely in your projects.

Now, let's dive into some practical examples of how external libraries and packages can supercharge your Python projects:

Web Development: Libraries like Flask and Django empower you to build web applications with ease. Flask, known for its simplicity, is an excellent choice for small to medium-sized projects, while Django offers a more comprehensive framework for larger applications.

Data Analysis and Visualization: If you're working with data, libraries like NumPy, pandas, and Matplotlib are indispensable. NumPy provides efficient numerical operations, pandas simplifies data manipulation, and Matplotlib enables you to create stunning visualizations.

Machine Learning: Python is a top choice for machine learning projects, thanks to libraries like scikit-learn, TensorFlow, and PyTorch. These libraries provide tools and algorithms for tasks such as classification, regression, clustering, and deep learning.

Game Development: Pygame is a popular library for creating 2D games in Python. It simplifies tasks like rendering graphics, handling user input, and managing game logic.

Natural Language Processing (NLP): Libraries like NLTK and spaCy offer robust tools for NLP tasks such as text tokenization, part-of-speech tagging, and sentiment analysis.

Database Interaction: If your project requires database access, libraries like SQLAlchemy make it straightforward to interact with relational databases, while MongoDB and PyMongo are a great combination for NoSQL databases.

Scientific Computing: For scientific and engineering applications, libraries like SciPy and SymPy provide a rich set of mathematical and scientific functions and capabilities.

IoT and Hardware Interaction: If you're working on IoT projects or hardware integration, libraries like Raspberry Pi's GPIO Zero and MicroPython offer support for interacting with sensors, motors, and other hardware components.

Game Development: Pygame is a popular library for creating 2D games in Python. It simplifies tasks like rendering graphics, handling user input, and managing game logic.

As you explore these libraries and packages, you'll notice that each one has its own documentation and community resources. This documentation is a valuable companion in your journey, offering detailed explanations, examples, and guides on how to use the library effectively.

Moreover, don't hesitate to delve into the source code of these libraries if you want to understand their inner workings or contribute to their development. Many open-source projects welcome contributions from the community, providing an excellent opportunity to learn and collaborate with fellow developers.

In summary, leveraging external libraries and packages is a game-changer in the world of Python development. It allows you to tap into a vast ecosystem of resources, enhance your productivity, and build powerful and feature-rich applications. So, don't hesitate to explore, experiment, and integrate these tools into your Python projects. Your journey as a Python developer is about to get even more exciting!

Welcome to the world of managing dependencies and version control in your Python projects. In this chapter, we'll explore the importance of keeping track of the external libraries and packages your project relies on, and how version control systems help maintain order and collaboration in your development journey.

Imagine you're building a complex Python application that relies on various libraries and packages, each contributing a unique set of functionalities. These external resources have their own release cycles, bug fixes, and updates, making it essential to manage them effectively.

Dependency management is the process of handling and tracking the external libraries and packages that your project depends on. It ensures that your application functions correctly with the specified versions of these dependencies. Without proper dependency management, your project can quickly become a tangled web of incompatible versions and unexpected behavior.

So, how do you manage dependencies in Python? The answer lies in a tool called **pip**. As you've seen before, **pip** allows you to install Python packages effortlessly. But it doesn't stop there; it also helps you create a **virtual environment**.

A virtual environment is a self-contained Python environment that isolates your project's dependencies from the system-wide Python installation. This isolation ensures that your project uses the specific versions of libraries and packages it requires, without interfering with other projects or the system Python.

Creating a virtual environment is as simple as running a single command:

pythonCopy code

```
python -m venv myenv
```

Here, **myenv** is the name of your virtual environment. You can replace it with any name you prefer.

Once your virtual environment is set up, you can **activate** it. Activating a virtual environment changes your system's Python interpreter to use the one from the virtual environment. This means that any packages you install while the environment is active will be isolated to that environment.

On Windows, you activate a virtual environment with:

pythonCopy code

```
myenv\Scripts\activate
```

On macOS and Linux, it's:

pythonCopy code

```
source myenv/bin/activate
```

With your virtual environment activated, you can now use **pip** to install project-specific dependencies. These dependencies are recorded in a file called **requirements.txt**. This file lists all the libraries and packages your project relies on, along with their specific versions.

To generate a **requirements.txt** file, you can use the **pip freeze** command:

pythonCopy code

```
pip freeze > requirements.txt
```

This command captures the currently installed packages and versions in your virtual environment and writes them to the **requirements.txt** file. Later, when you or someone else wants to recreate the same environment, they can do so by running:

pythonCopy code

pip install -r requirements.txt

This ensures that the project's dependencies remain consistent across different environments and over time.

Now, let's shift our focus to **version control**. Version control is the practice of tracking and managing changes to your codebase over time. It provides a historical record of your project, allowing you to collaborate with others, revert to previous states, and manage complex development workflows.

One of the most popular version control systems is **Git**. Git is an open-source distributed version control system that empowers developers to work collaboratively on code. It tracks changes in your project's source code, records a history of these changes, and allows you to synchronize your code with remote repositories.

With Git, you can create a **repository**, which is essentially a container for your project's code and version history. Git repositories can be hosted on various platforms, such as GitHub, GitLab, or Bitbucket, providing a central location for collaboration and code sharing.

To start using Git, you need to initialize a Git repository in your project's directory:

pythonCopy code

```
git init
```

This command initializes an empty Git repository in the current directory. Git will start tracking changes to your files from this point forward.

Once you've made some changes to your project, you can use Git to **commit** these changes to the repository. A commit is a snapshot of your code at a particular point in time, along with a descriptive message explaining the changes you've made.

pythonCopy code

git add . git commit -m "Initial commit"

In this example, we first use **git add .** to stage all changes in the current directory for the commit. Then, we commit these changes with a descriptive message.

Git's ability to create branches is a powerful feature. A **branch** is essentially a separate line of development that allows you to work on new features, bug fixes, or experiments without affecting the main codebase. You can create branches, switch between them, and merge them when your work is complete.

To create a new branch, you can use:

pythonCopy code

```
git branch new_feature
```

This command creates a new branch named **new_feature**. You can switch to this branch with:

pythonCopy code

```
git checkout new_feature
```

While working on your new feature or bug fix, you can make commits as usual. When your work is complete, you can merge it back into the main codebase by switching to the main branch and using **git merge**.

pythonCopy code

```
git checkout main git merge new_feature
```

Git's branching and merging capabilities facilitate collaborative development, allowing multiple developers to work on different parts of a project simultaneously. It also helps in maintaining a clean and organized codebase.

Collaboration is a fundamental aspect of software development, and Git makes it easy to work with others. You can **clone** remote repositories to your local machine, make changes, and push those changes back to the remote repository for others to see and review.

To clone a remote repository, you can use the **git clone** command:

pythonCopy code

```
git clone https://github.com/username/repository.git
```

This command creates a copy of the remote repository on your local machine, allowing you to work on it locally.

When you're ready to share your changes with others, you can **push** them to the remote repository. Others can then **pull** your changes to their own local copies.

pythonCopy code

```
git push origin main
```

This command pushes your local **main** branch to the remote repository named **origin**.

Version control systems like Git also provide features for resolving conflicts when multiple developers make changes to the same part of the code simultaneously. Git helps identify these conflicts, allowing you to review and manually resolve them.

In summary, managing dependencies and version control are essential skills for any developer, whether you're working on small personal projects or collaborating with a large team on a complex software project. Proper dependency management ensures that your project remains stable and consistent, while version control systems like Git empower you to track changes, collaborate effectively, and maintain a clean and organized codebase.

As you continue your journey in Python development, these practices will become second nature, helping you become a more efficient and collaborative programmer. So, embrace the power of dependency management and version control, and you'll find that they are invaluable tools in your toolbox. Happy coding!

Chapter 5: Advanced File Handling and Manipulation

Welcome to the exciting world of file input/output (I/O) techniques in Python that go beyond the basics. In this chapter, we'll delve into advanced file handling methods and explore how you can interact with files, directories, and complex data formats to take your Python projects to the next level.

You're already familiar with the fundamentals of file I/O in Python, where you can open, read, and write to files using built-in functions like **open()**, **read()**, and **write()**. These basic techniques serve you well for simple tasks, but what if your project demands more advanced file manipulation?

Let's start by exploring how to work with files in different modes. In Python, the **open()** function allows you to specify a mode when opening a file. The default mode is 'r' (read), but there are several other modes you can use to suit your needs.

For instance, if you want to open a file for writing, you can use 'w' mode:

pythonCopy code

```
with open('example.txt', 'w') as file: file.write('Hello, world!')
```

In this example, the 'w' mode opens the file 'example.txt' for writing. If the file already exists, it's truncated (i.e., its contents are erased). If it doesn't exist, a new file is created. The **with** statement ensures that the file is properly closed when you're done with it.

But what if you want to add content to an existing file without erasing its current contents? You can achieve that with 'a' mode (append):

pythonCopy code

```
with        open('example.txt',        'a')        as        file:
file.write('\nAppending some more text.')
```

Here, we're using 'a' mode to append new content to the existing 'example.txt' file. The '\n' character inserts a line break, ensuring that the new content starts on a new line.

For even more control over file I/O, you can use the 'b' mode to work with binary files. Binary mode is particularly useful when dealing with non-textual data, such as images, audio, or binary files like SQLite databases.

pythonCopy code

```
with  open('binary_data.bin', 'rb') as binary_file: data =
binary_file.read()
```

In this snippet, 'rb' mode opens 'binary_data.bin' in binary read mode, allowing you to read its binary content. Similarly, 'wb' mode is used for binary write operations.

Now, let's venture into more advanced file I/O techniques, starting with working with directories and file paths. Python's **os** module provides a wealth of functions for directory and file manipulation.

You can use **os.mkdir()** to create a new directory:

pythonCopy code

```
import os os.mkdir('new_directory')
```

This creates a directory named 'new_directory' in the current working directory. To check if a directory exists, you can use **os.path.exists()**:

pythonCopy code

```
if  os.path.exists('new_directory'):  print('The  directory
exists.') else: print('The directory does not exist.')
```

os.path.exists() returns **True** if the specified path (directory or file) exists, and **False** otherwise.

When dealing with file paths, Python's **os.path** module helps you handle them in a platform-independent way. The

121

os.path.join() function concatenates directory and file names to create a path:

pythonCopy code

```
import os file_path = os.path.join('my_directory', 'myfile.txt')
```

The resulting **file_path** will contain the correct path, whether you're on a Windows or Unix-based system.

Now, let's explore working with structured data formats, such as JSON and CSV files. These formats are commonly used for data exchange and storage.

Python's **json** module makes it easy to work with JSON data. You can use **json.dump()** to write Python data structures to a JSON file:

pythonCopy code

```
import json data = {'name': 'John', 'age': 30, 'city': 'New York'} with open('data.json', 'w') as json_file: json.dump(data, json_file)
```

This code snippet writes the Python dictionary **data** to a JSON file named 'data.json'. Later, you can read the JSON data back into Python using **json.load()**:

pythonCopy code

```
with open('data.json', 'r') as json_file: loaded_data = json.load(json_file) print(loaded_data)
```

The **loaded_data** variable now contains the dictionary data loaded from 'data.json'.

When it comes to CSV (Comma-Separated Values) files, Python's **csv** module simplifies the process of reading and writing tabular data.

pythonCopy code

```
import csv # Writing data to a CSV file data = [['Name', 'Age'], ['John', 30], ['Alice', 25], ['Bob', 35]] with
```

open('data.csv', 'w', newline='') as csv_file: writer = csv.writer(csv_file) writer.writerows(data)

Here, we're creating a CSV file named 'data.csv' and writing a list of lists (**data**) to it. The **newline=''** argument ensures that newline characters are handled correctly in different operating systems.

To read data from a CSV file, you can use the **csv.reader()**:

pythonCopy code

with open('data.csv', 'r') as csv_file: reader = csv.reader(csv_file) for row in reader: print(row)

The **csv.reader()** allows you to iterate through the rows of the CSV file and access the data.

Sometimes, you may encounter more complex data formats, such as XML or Excel spreadsheets. Python provides libraries like **xml.etree.ElementTree** and **openpyxl** to work with these formats, allowing you to extract, manipulate, and generate structured data efficiently.

In this chapter, we've explored advanced file I/O techniques that go beyond the basics, including working with different file modes, managing directories and file paths, and handling structured data formats like JSON and CSV. These skills will empower you to tackle more complex data processing tasks and build robust Python applications that interact seamlessly with files and directories.

As you continue your journey as a Python developer, you'll find these advanced file I/O techniques to be valuable tools in your toolkit, enabling you to handle diverse data sources and create more sophisticated applications. So, don't hesitate to dive deeper into these topics, experiment, and expand your Python horizons. Happy coding!

Welcome to the fascinating realm of manipulating data streams and binary files in Python. In this chapter, we'll embark on a journey to explore how Python handles data at a lower level, diving into binary files, byte manipulation, and stream processing. By the end of this chapter, you'll have a deep understanding of how to work with binary data and streams effectively.

In the world of programming, data comes in many forms, but at its core, it often boils down to ones and zeros, the building blocks of digital information. Binary files, unlike text files, don't use character encodings like UTF-8 or ASCII; instead, they store data in its raw binary form.

Binary files can contain anything from images and audio files to database files and more. Python provides various tools and modules to read, write, and manipulate these binary files, allowing you to unlock their potential.

Let's begin by discussing how to read data from a binary file. To read binary data in Python, you can use the built-in **open()** function with the 'rb' mode (read binary). For instance, suppose you have a binary file named 'data.bin':

pythonCopy code

```
with open('data.bin', 'rb') as binary_file: binary_data = binary_file.read()
```

In this code snippet, 'data.bin' is opened in 'rb' mode, allowing you to read its binary content. The **binary_file.read()** method reads the entire contents of the binary file into the **binary_data** variable.

Once you have the binary data in memory, you can manipulate it in various ways. For example, you can iterate through the bytes, perform bitwise operations, or convert it to other data types.

Binary data is often used in scenarios where efficiency and compactness are crucial, such as when dealing with large datasets or low-level hardware interactions. Python's ability to work with binary data gives you the flexibility to handle such situations effectively.

Now, let's explore the concept of data streams. A data stream is a sequence of data elements made available over time. In Python, streams are typically used for input and output operations, allowing you to read and write data continuously, rather than all at once.

The **io** module in Python provides a foundation for working with data streams. You can create stream objects to read or write data incrementally, making it ideal for tasks like parsing large log files or processing real-time data streams.

Let's take a look at reading data from a stream using the **io** module. Suppose you have a stream of data from a network socket or a file-like object:

```python
import io
# Simulate a data stream
data_stream = io.BytesIO(b"Hello, world!")
# Read data from the stream
data = data_stream.read()
```

In this example, we use **io.BytesIO** to create an in-memory binary data stream with the initial content "Hello, world!". We then read the data from the stream into the **data** variable. This concept of data streams is invaluable when dealing with continuous data sources or when you want to avoid loading the entire dataset into memory.

On the flip side, writing data to a stream is equally straightforward. You can create a writable stream object and use it to write data incrementally:

```python
import io
# Create a writable data stream
output_stream = io.BytesIO()
# Write data to the stream
output_stream.write(b"Hello,")
output_stream.write(b"
```

world!") # Retrieve the content from the stream
output_data = output_stream.getvalue()

Here, we create an **io.BytesIO** stream, write "Hello," and "
world!" to it separately, and then retrieve the combined
content using **getvalue()**. This approach is incredibly useful
for generating dynamic or large datasets without consuming
excessive memory.

Binary files, data streams, and byte manipulation often come
together when you need to perform tasks like encoding and
decoding data. Python provides various encoding and
decoding schemes to convert binary data into text or other
formats.

For example, you can encode binary data using Base64
encoding, which converts binary data into a plain text format
that is safe for transmission or storage. Python's **base64**
module makes encoding and decoding a breeze:

pythonCopy code

```
import base64 # Encode binary data to Base64 binary_data
=    b"Hello,    world!"    encoded_data    =
base64.b64encode(binary_data) # Decode Base64 data back
to    binary    decoded_data    =
base64.b64decode(encoded_data)
```

In this snippet, **base64.b64encode()** encodes the binary data
into Base64, and **base64.b64decode()** decodes it back to its
original binary form. Base64 encoding is commonly used
when working with data over the internet, like email
attachments or transmitting binary data via HTTP.

Another crucial concept to grasp when working with binary
data is endianness. Endianness refers to the byte order in
which multibyte data types are stored in memory. Python
provides the **struct** module to help you handle endianness
and pack/unpack binary data efficiently.

Consider a scenario where you need to deal with binary data that follows a specific format, such as a network packet header. Here's how you can use the **struct** module to pack and unpack binary data:

pythonCopy code

```
import struct # Define a binary format with little-endian byte order binary_format = '<I 2s' # Pack data into binary format packed_data = struct.pack(binary_format, 42, b"Hi") # Unpack binary data unpacked_data = struct.unpack(binary_format, packed_data)
```

In this example, the '<' character in **binary_format** specifies little-endian byte order, which is common in many architectures. We pack two values, an integer (42) and a 2-character string (b"Hi"), into the binary format and then unpack them to retrieve the original data.

Understanding binary data, data streams, and byte manipulation opens up a world of possibilities in Python. Whether you're working with low-level hardware communication, network protocols, or file formats that demand precise control over binary data, Python equips you with the tools and libraries to handle these challenges effectively.

As you continue your journey as a Python developer, these skills will empower you to tackle more complex data manipulation tasks, making you a more versatile and capable programmer. So, embrace the world of binary data and data streams; you'll find them to be powerful allies in your programming adventures. Happy coding!

Chapter 6: Web Automation and Interaction

Welcome to the exciting world of automating web tasks with Python. In this chapter, we'll embark on a journey that will empower you to harness the full potential of Python for automating repetitive and time-consuming web-related tasks. Imagine having the ability to automate web scraping, form filling, interaction with web APIs, and much more, all at your fingertips.

Python has become a powerhouse for web automation, thanks to a rich ecosystem of libraries and tools. Whether you're a developer, data scientist, or someone looking to streamline daily web activities, Python's versatility in this domain is a valuable asset.

Let's kick things off by exploring web scraping, a fundamental aspect of web automation. Web scraping involves extracting data from websites, making it available for analysis or further processing. Python offers several libraries, such as BeautifulSoup and Scrapy, that simplify the process of scraping web content.

Consider a scenario where you need to extract information from a news website. With Python and BeautifulSoup, you can create a web scraper to fetch headlines, article summaries, and publication dates. Here's a glimpse of how it's done:

pythonCopy code

```
import requests from bs4 import BeautifulSoup # Send a
GET request to the news website url = 'https://example-
news-website.com' response = requests.get(url) # Parse the
HTML content of the page soup =
BeautifulSoup(response.text, 'html.parser') # Extract
headlines, summaries, and publication dates headlines =
```

```
soup.find_all('h2', class_='headline') summaries =
soup.find_all('p', class_='summary') dates =
soup.find_all('span', class_='date') # Process and use the
```
extracted data as needed

In this code snippet, we use the **requests** library to send a GET request to the news website, fetch the HTML content, and then parse it with BeautifulSoup. We identify and extract the desired elements, such as headlines, summaries, and publication dates, which can be further processed or stored for analysis.

Web scraping is a powerful tool not only for data collection but also for monitoring websites, tracking changes, and staying up to date with relevant information.

Moving on, web automation is not limited to scraping data; it extends to interacting with websites and web applications. Consider a scenario where you need to automate the process of filling out forms on a website. This can be a tedious and repetitive task, but Python can simplify it for you.

Python's Selenium library is your ally in automating web interactions. Selenium allows you to control a web browser programmatically, enabling actions like clicking buttons, filling out forms, and navigating through web pages. Let's look at a simple example of how you can automate form filling with Selenium:

pythonCopy code

```
from selenium import webdriver # Create a web driver
(e.g., Chrome) driver =
webdriver.Chrome(executable_path='/path/to/chromedrive
r') # Navigate to a webpage with a form
driver.get('https://example-form-website.com') # Find the
form elements and fill them out username_input =
```

```
driver.find_element_by_id('username') password_input =
driver.find_element_by_id('password')
```
username_input.send_keys('your_username')
password_input.send_keys('your_password') # Submit the
form submit_button = driver.find_element_by_id('submit-
button') submit_button.click() # Close the browser window
driver.quit()

In this script, we use Selenium to automate the process of
filling out a form on a website. We locate the form elements
by their IDs and use the **send_keys()** method to input data.
After filling out the form, we submit it by locating the submit
button and clicking it. Finally, we close the browser window.

Web automation is not limited to interacting with websites
alone; it extends to working with web APIs. Web APIs
(Application Programming Interfaces) allow you to interact
with online services and retrieve data in a structured format,
typically JSON or XML.

Python's **requests** library is a go-to choice for making HTTP
requests to web APIs. Whether you want to fetch weather
data, access financial information, or integrate with social
media platforms, Python's ability to handle web APIs
simplifies the process.

For instance, let's say you want to retrieve weather data
from a weather API:

pythonCopy code

import requests # Define the API endpoint and parameters
api_url = 'https://api.example-weather-website.com'
params = {'location': 'New York', 'units': 'metric'} # Send a
GET request to the API response = requests.get(api_url,
params=params) # Parse the JSON response weather_data
= response.json() # Extract and use weather information

```python
temperature = weather_data['main']['temp'] description =
weather_data['weather'][0]['description']  # Process and
display the weather information
```

In this example, we define the API endpoint and parameters, send a GET request using **requests**, and then parse the JSON response to extract weather data. This data can be utilized for various purposes, such as displaying it on a website or integrating it into a mobile app.

The automation possibilities with Python are virtually limitless when it comes to the web. You can automate tasks like web testing, content publishing, data analysis, and more. Python's web automation capabilities open up a world of efficiency and creativity, allowing you to save time and focus on higher-level tasks.

As you continue your exploration of web automation with Python, you'll find that it's a valuable skill that can significantly enhance your productivity and simplify complex web-related tasks. So, dive in, experiment, and let Python be your companion in the exciting journey of web automation. Happy automating!

Welcome to the captivating world of interacting with websites and web services using Python. In this chapter, we're going to embark on a journey that will empower you to navigate the vast landscape of web development, from making HTTP requests to integrating with external APIs. So, fasten your seatbelts, because we're about to explore how Python becomes your trusty companion in this exciting adventure.

At the heart of web interaction lies the HTTP protocol, the backbone of the World Wide Web. HTTP, short for Hypertext Transfer Protocol, is the communication standard that allows your web browser to fetch web pages, send form data, and retrieve resources like images and scripts. Python provides powerful libraries to work with HTTP, making it a breeze to send requests and process responses.

The **requests** library is your Swiss Army knife for making HTTP requests in Python. Whether you want to retrieve web pages, send data to a server, or interact with RESTful APIs, **requests** simplifies the process. Let's dive into an example to see how straightforward it can be:

pythonCopy code

```python
import requests # Define the URL you want to retrieve url = 'https://www.example-website.com' # Send a GET request to the URL response = requests.get(url) # Check if the request was successful (status code 200) if response.status_code == 200: # Print the content of the response print(response.text) else: # Handle the error print('Error:', response.status_code)
```

In this code snippet, we use **requests** to send a GET request to a website and retrieve its content. If the request is successful (indicated by a status code of 200), we print the content. Otherwise, we handle the error gracefully. With just a few lines of code, you can fetch web pages and start working with their content.

Now, let's venture into the realm of web APIs (Application Programming Interfaces). Web APIs allow you to access data and functionality provided by external services over the internet. Python's ability to interact with APIs makes it a versatile tool for integrating with a wide range of services, from social media platforms to weather data providers.

Consider a scenario where you want to retrieve data from a public API, like the OpenWeatherMap API, to get the current weather conditions for a specific location. Python, with its **requests** library, simplifies the process:

pythonCopy code

```python
import requests # Define the API endpoint and parameters
api_url = 'https://api.openweathermap.org/data/2.5/weather'
params = { 'q': 'New York', 'appid': 'your_api_key', } # Send a GET request to the API
response = requests.get(api_url, params=params) # Check if the request was successful (status code 200)
if response.status_code == 200: # Parse the JSON response
    weather_data = response.json() # Extract and use weather information
    temperature = weather_data['main']['temp']
    description = weather_data['weather'][0]['description'] # Display the weather information
    print(f'Temperature: {temperature}°C')
    print(f'Condition: {description}')
else: # Handle the error
    print('Error:', response.status_code)
```

In this example, we define the API endpoint and parameters, send a GET request to the OpenWeatherMap API, and parse the JSON response to extract weather data. This data can then be displayed or further processed as needed.

Web APIs open up a world of possibilities for developers, allowing you to tap into a vast array of services and data sources to enrich your applications or automate tasks. Whether you're building a weather app, fetching stock market data, or integrating with social media platforms, Python's simplicity and power shine in this domain.

Moving forward, web interaction often involves not just sending data but also receiving it in structured formats like JSON or XML. Python's built-in libraries make it effortless to parse and work with such data.

For instance, let's explore parsing JSON, a common data format used by many web services:

pythonCopy code

```python
import json # Sample JSON data json_data = '{"name":
"John", "age": 30, "city": "New York"}' # Parse JSON into a
Python dictionary data_dict = json.loads(json_data) # Access
and use the data name = data_dict['name'] age =
data_dict['age'] city = data_dict['city'] # Display the parsed
data print(f'Name: {name}') print(f'Age: {age}')
print(f'City: {city}')
```

In this snippet, we use the **json.loads()** method to parse
JSON data into a Python dictionary, making it easy to access
and use the structured information. Whether you're dealing
with JSON, XML, or other data formats, Python provides the
tools to handle them seamlessly.

Python's versatility extends to web authentication as well.
Many web services and APIs require authentication to
ensure secure access. Python simplifies the process with
libraries like **requests**, which allow you to include
authentication credentials in your requests.

Let's look at an example of using basic authentication with
the **requests** library:

pythonCopy code

```python
import requests from requests.auth import HTTPBasicAuth
# Define the URL and authentication credentials url =
'https://example-api.com/data' username =
'your_username' password = 'your_password' # Send a GET
request with basic authentication response =
requests.get(url, auth=HTTPBasicAuth(username,
password)) # Check if the request was successful (status
code 200) if response.status_code == 200: # Process the
response data data = response.json() # Handle the data as
needed else: # Handle the error print('Error:',
response.status_code)
```

In this code, we import **HTTPBasicAuth** from **requests.auth** and include it in the request using the **auth** parameter. This way, you can securely access web services that require basic authentication.

As you continue your journey into web interaction with Python, you'll discover countless possibilities, from building web applications to automating complex tasks. Python's elegance and simplicity empower you to harness the full potential of the web, making it a versatile tool for developers and data enthusiasts alike.

So, embrace the world of web interaction with Python, and let it be your guide in exploring new horizons, building exciting applications, and simplifying the way you interact with websites and web services. Happy coding!

Chapter 7: Database Connectivity and Manipulation

Welcome to the fascinating realm of connecting to databases with Python. In this chapter, we're embarking on a journey that will equip you with the skills to seamlessly interact with databases, whether you're working with traditional relational databases, NoSQL databases, or even connecting to remote database servers. So, let's dive into the world of data storage and retrieval using Python.

Databases play a pivotal role in modern software development. They serve as structured repositories for storing and retrieving data, making them an integral part of applications, from web applications to data analysis pipelines. Python offers a robust ecosystem of libraries and tools that make it a breeze to connect to and manipulate databases.

To start our journey, let's explore how Python connects to relational databases, a category that includes widely used systems like MySQL, PostgreSQL, and SQLite. Python's standard library includes a module called **sqlite3**, which provides a straightforward way to interact with SQLite databases, a popular choice for embedded and lightweight database needs.

Imagine you're working on a project that requires a simple local database to store user information. Python's **sqlite3** module can be your ally in this endeavor. Let's take a look at how to create and interact with an SQLite database:

pythonCopy code

import sqlite3 # Connect to an SQLite database (or create one if it doesn't exist) conn = sqlite3.connect('user_data.db') # Create a cursor object to interact with the database cursor = conn.cursor() # Create a table to store user information cursor.execute("' CREATE TABLE IF NOT EXISTS users (id INTEGER PRIMARY KEY, username TEXT NOT NULL, email TEXT NOT NULL) "') # Insert data into the table cursor.execute('INSERT INTO users (username, email) VALUES (?, ?)', ('john_doe', 'john@example.com')) # Commit the changes and close the connection conn.commit() conn.close()

In this code, we connect to an SQLite database using **sqlite3.connect()**, create a cursor to execute SQL commands, and then define a table schema for storing user information. We insert a user record into the table and commit the changes.

Python's **sqlite3** module is well-suited for small to medium-sized applications and prototyping, thanks to its simplicity and lightweight nature. However, when working with more robust relational databases like MySQL or PostgreSQL, Python provides dedicated libraries like **mysql-connector-python** and **psycopg2** to facilitate the connection and data manipulation.

Speaking of data manipulation, SQL (Structured Query Language) is the language of choice for interacting with relational databases. Python's database libraries allow you to execute SQL queries directly from your Python code, giving you the flexibility to retrieve, modify, and manage data as needed.

Let's consider a scenario where you want to retrieve user information from a MySQL database. Using the **mysql-connector-python** library, you can establish a connection and execute SQL queries like this:

pythonCopy code

```python
import mysql.connector # Establish a connection to the
MySQL database conn = mysql.connector.connect(
host='localhost',                       user='your_username',
password='your_password', database='your_database' ) #
Create a cursor for executing SQL queries cursor =
conn.cursor() # Execute a SELECT query to retrieve user data
cursor.execute('SELECT username, email FROM users') #
Fetch all the results users = cursor.fetchall() # Iterate
through the results and process the data for user in users:
username, email = user print(f'Username: {username},
Email: {email}') # Close the cursor and the connection
cursor.close() conn.close()
```

In this example, we connect to a MySQL database, execute a
SELECT query to retrieve user data, and then process and
display the results.

Moving beyond traditional relational databases, Python's
versatility extends to NoSQL databases like MongoDB.
NoSQL databases are known for their flexibility and
scalability, making them ideal for handling large volumes of
unstructured data.

Python provides libraries like **pymongo** that simplify the
process of interacting with MongoDB. Imagine you're
building a web application that needs to store and retrieve
user profiles in a MongoDB database. Let's see how Python
can assist you in this endeavor:

pythonCopy code

```python
import pymongo # Establish a connection to the MongoDB
server                        client                        =
pymongo.MongoClient('mongodb://localhost:27017/')    #
Access the database db = client['user_profiles'] # Access the
collection (analogous to a table in SQL) collection =
```

```
db['users']  # Insert a user profile into the collection
user_profile = { 'username': 'jane_doe', 'email':
'jane@example.com', 'age': 28 } insert_result =
collection.insert_one(user_profile) # Retrieve and display
the inserted document's ID print(f'Inserted document ID:
{insert_result.inserted_id}') # Query the collection for user
profiles query = {'age': {'$gte': 25}} matching_profiles =
collection.find(query) # Iterate through the results and
process the data for profile in matching_profiles:
print(f'Username: {profile["username"]}, Email:
{profile["email"]}, Age: {profile["age"]}') # Close the
connection client.close()
```

In this example, we use **pymongo** to connect to a MongoDB server, access a database and collection, and perform operations like inserting and querying user profiles.

Python's adaptability also extends to working with remote databases hosted on cloud platforms like Amazon Web Services (AWS) and Google Cloud Platform (GCP). These platforms offer managed database services that enable you to focus on your application logic while leaving the database management to the cloud provider. Python libraries like **boto3** for AWS and **google-cloud-firestore** for GCP make it straightforward to connect to cloud-hosted databases.

As you continue your journey into the world of database connectivity with Python, you'll discover that Python's rich ecosystem of libraries and its ease of use make it an excellent choice for a wide range of database-related tasks. Whether you're building web applications, data analysis pipelines, or IoT projects, Python's database connectivity capabilities will empower you to efficiently store and retrieve data, opening up endless possibilities for your projects.

So, embrace the power of Python in connecting to databases, and let it be your trusted companion in your data-driven endeavors. Happy coding!

Welcome to the exciting world of SQL querying and data manipulation with Python! In this chapter, we're embarking on a journey that will equip you with the skills to harness the power of SQL, the language of databases, and seamlessly integrate it with Python for efficient data retrieval, transformation, and manipulation. So, let's dive into the art of querying and shaping data using Python and SQL.

SQL, which stands for Structured Query Language, is the universal language of databases. Whether you're working with relational databases like MySQL, PostgreSQL, SQLite, or non-relational databases like MongoDB, SQL provides a standardized way to interact with data. Python's extensive ecosystem of database libraries and connectors makes it an ideal companion for SQL-based tasks.

Let's start by exploring the fundamental concepts of SQL querying. Imagine you have a database that stores information about books, including their titles, authors, and publication years. You can use SQL to retrieve specific data from this database.

Here's a simple SQL query written in Python to fetch all the books in the database:

pythonCopy code

```
import sqlite3 # Connect to the SQLite database conn = sqlite3.connect('library.db') # Create a cursor object to execute SQL commands cursor = conn.cursor() # Define an SQL query to select all books query = 'SELECT * FROM books' # Execute the query and fetch the results cursor.execute(query) books = cursor.fetchall() # Close the cursor and the connection cursor.close() conn.close() # Print the retrieved books for book in books: print(book)
```

In this example, we use the **sqlite3** library to connect to an SQLite database, create a cursor for executing SQL queries, and then execute a simple SQL **SELECT** query to retrieve all the books from the "books" table. The results are fetched and printed to the console.

SQL queries can be as simple or as complex as your data retrieval needs. You can use SQL to filter data, aggregate values, join tables, and perform a wide range of operations. Python's SQL libraries make it easy to incorporate these queries into your applications, allowing you to retrieve exactly the data you need.

Let's dive a bit deeper into SQL querying by considering a more advanced scenario. Imagine you want to find all books published after the year 2000 and written by a specific author. Here's how you can accomplish this using SQL and Python:

pythonCopy code

import sqlite3 # Connect to the SQLite database conn = sqlite3.connect('library.db') # Create a cursor object to execute SQL commands cursor = conn.cursor() # Define the author's name and the publication year author_name = 'Jane Doe' publication_year = 2000 # Define an SQL query to select books by the specified author published after 2000 query = 'SELECT title FROM books WHERE author = ? AND publication_year > ?' # Execute the query with parameters cursor.execute(query, (author_name, publication_year)) # Fetch the results books = cursor.fetchall() # Close the cursor and the connection cursor.close() conn.close() # Print the retrieved books for book in books: print(book[0])

In this example, we use SQL's **WHERE** clause to filter the results based on the author's name and the publication year.

We also use parameterized queries to safely pass values into the SQL statement.

Python's SQL libraries support various database systems, so you can apply the same SQL skills to different databases seamlessly. Whether you're querying data from a MySQL database, a PostgreSQL database, or any other database, the process remains consistent.

Beyond querying, Python enables you to manipulate and transform data with ease. For instance, you might want to calculate the total number of books written by each author in your library database. Here's how you can achieve this by combining SQL querying with Python's data processing capabilities:

pythonCopy code

```
import sqlite3 # Connect to the SQLite database conn = sqlite3.connect('library.db') # Create a cursor object to execute SQL commands cursor = conn.cursor() # Define an SQL query to count books per author query = ''' SELECT author, COUNT(*) as total_books FROM books GROUP BY author ''' # Execute the query cursor.execute(query) # Fetch the results author_book_counts = cursor.fetchall() # Close the cursor and the connection cursor.close() conn.close() # Print the author and their total number of books for author, total_books in author_book_counts: print(f'Author: {author}, Total Books: {total_books}')
```

In this example, we use SQL's **GROUP BY** clause to group books by author and then count the number of books per author. Python processes the results, making it easy to display the author's name and their corresponding total number of books.

Python's flexibility extends to data transformation and visualization. You can use popular libraries like Pandas and

Matplotlib to further analyze and visualize your data. For instance, you could create a bar chart to visualize the number of books per author.

pythonCopy code

```
import sqlite3 import pandas as pd import matplotlib.pyplot as plt # Connect to the SQLite database conn = sqlite3.connect('library.db') # Query the database and load the results into a Pandas DataFrame query = ''' SELECT author, COUNT(*) as total_books FROM books GROUP BY author ''' df = pd.read_sql_query(query, conn) # Close the connection conn.close() # Create a bar chart to visualize the data plt.figure(figsize=(10, 6)) plt.bar(df['author'], df['total_books']) plt.xlabel('Author') plt.ylabel('Total Books') plt.title('Number of Books per Author') plt.xticks(rotation=90) plt.tight_layout() # Display the chart plt.show()
```

In this example, we first query the database using SQL, load the results into a Pandas DataFrame, and then use Matplotlib to create a bar chart. Python's data manipulation and visualization libraries provide a powerful toolkit for exploring and presenting your data.

As you continue your journey in SQL querying and data manipulation with Python, you'll find that the combination of these two powerful tools opens up a world of possibilities. Whether you're building data-driven applications, conducting data analysis, or creating informative visualizations, Python's versatility and SQL's querying capabilities will be your trusted companions on your data adventure. So, embrace the power of Python and SQL, and let them guide you as you navigate the vast landscape of data manipulation and analysis. Happy querying and coding!

Chapter 8: Multithreading and Parallel Processing

Welcome to the fascinating world of multithreading in Python! In this chapter, we're diving into a topic that can significantly enhance the performance of your Python applications by allowing them to execute multiple tasks concurrently. Multithreading is like having multiple workers in a factory, each handling a different task, and in Python, it's a valuable tool for achieving parallelism.

Python, known for its simplicity and versatility, provides several ways to work with threads. Understanding how to leverage multithreading effectively can be a game-changer for applications that need to perform multiple tasks simultaneously, such as web scraping, data processing, and handling concurrent user requests in a web server.

At its core, a thread is a lightweight, independent unit of execution within a process. Think of a process as the overall program, and threads as individual workers within that program. These threads share the same memory space, making communication and data sharing between them efficient.

To get started with multithreading in Python, you can use the built-in **threading** module. This module allows you to create and manage threads effortlessly. Let's begin with a simple example.

Imagine you have a task that involves fetching data from multiple websites. Traditionally, without multithreading, you'd have to fetch data from one website at a time, which can be slow if you have a long list of websites to process. With multithreading, you can fetch data from multiple websites simultaneously, greatly speeding up the process.

Here's a basic Python script that demonstrates multithreading for fetching data from multiple websites:

```
pythonCopy code
import threading import requests # Define a list of websites
to fetch data from websites = ['https://example.com',
'https://python.org', 'https://github.com'] # Function to
fetch data from a website def fetch_website_data(url):
response = requests.get(url) print(f"Fetched data from {url},
Length: {len(response.text)}") # Create thread objects for
each website threads = [] for website in websites: thread =
threading.Thread(target=fetch_website_data,
args=(website,)) threads.append(thread) # Start the threads
for thread in threads: thread.start() # Wait for all threads to
finish for thread in threads: thread.join() print("All threads
have finished.")
```

In this script, we import the **threading** module and define a
list of websites to fetch data from. We then create a function
fetch_website_data that takes a URL as an argument and
uses the **requests** library to fetch data from that URL. Next,
we create thread objects for each website, passing the
fetch_website_data function as the target.

By starting these threads and waiting for them to finish using
the **start** and **join** methods, respectively, we enable
concurrent fetching of data from multiple websites. This
approach significantly reduces the time it takes to complete
the task compared to fetching data sequentially.

However, it's important to note that Python's Global
Interpreter Lock (GIL) can limit the performance benefits of
multithreading in CPU-bound tasks. The GIL restricts the
execution of multiple threads in a single Python process, so
in CPU-bound scenarios, multithreading might not provide
the expected speedup. In such cases, multiprocessing or
asynchronous programming may be more suitable.

For I/O-bound tasks, like the web scraping example above, multithreading is effective because the GIL is released during I/O operations, allowing other threads to run. But for CPU-bound tasks, where computation is the bottleneck, you might want to explore Python's multiprocessing module or asynchronous libraries like asyncio.

Python's **multiprocessing** module allows you to create multiple processes, each with its own Python interpreter and memory space. This approach is suitable for CPU-bound tasks since it leverages multiple CPU cores and can provide significant speed improvements.

Let's consider an example where we calculate the factorial of multiple numbers concurrently using the **multiprocessing** module:

pythonCopy code

```
import multiprocessing # Function to calculate factorial of a number def calculate_factorial(num): result = 1 for i in range(1, num + 1): result *= i return result # List of numbers to calculate factorial for numbers = [5, 10, 15, 20] # Create a pool of processes pool = multiprocessing.Pool() # Map the function to the numbers and retrieve results results = pool.map(calculate_factorial, numbers) # Close the pool pool.close() pool.join() # Print the results for num, result in zip(numbers, results): print(f"Factorial of {num} is {result}")
```

In this example, we use the **multiprocessing.Pool** class to create a pool of processes, and then we use the **map** method to distribute the calculation of factorials for multiple numbers across these processes. This approach takes advantage of multiple CPU cores and can significantly speed up CPU-bound computations.

Asynchronous programming is another powerful way to achieve concurrency in Python, especially for I/O-bound tasks that involve waiting for external resources, such as network requests or file operations. Python's **asyncio** library provides the tools to write asynchronous code using **async** and **await** keywords.

Here's a simple example of asynchronous web scraping using **asyncio** and the **aiohttp** library:

pythonCopy code

```
import asyncio import aiohttp # List of URLs to fetch data from asynchronously urls = ['https://example.com', 'https://python.org', 'https://github.com'] async def fetch_website_data(url): async with aiohttp.ClientSession() as session: async with session.get(url) as response: data = await response.text() print(f"Fetched data from {url}, Length: {len(data)}") # Create an event loop and gather tasks for fetching data async def main(): tasks = [fetch_website_data(url) for url in urls] await asyncio.gather(*tasks) # Run the event loop if __name__ == '__main__': asyncio.run(main())
```

In this example, we define an asynchronous function **fetch_website_data** that fetches data from a URL using **aiohttp**. We create an event loop and gather tasks for fetching data from multiple URLs concurrently. The **asyncio.run(main())** line starts the event loop.

Asynchronous programming is particularly well-suited for scenarios where you have many I/O-bound operations, as it allows you to efficiently manage multiple tasks without the overhead of creating and managing threads or processes.

In summary, understanding multithreading in Python opens up new possibilities for concurrent execution of tasks, whether you're dealing with I/O-bound or CPU-bound workloads. Python provides various tools and libraries, such as **threading, multiprocessing**, and **asyncio**, to help you achieve concurrency based on the specific requirements of your application.

By selecting the right approach and tools for your task, you can harness the power of multithreading and concurrent programming in Python to build more responsive and efficient applications. So, go ahead and explore the world of concurrency, and let Python be your guide to achieving parallelism and boosting performance in your projects. Happy coding!

Welcome to the exciting world of achieving parallelism in your Python applications for improved performance! In this chapter, we'll delve into the art of parallel programming, a technique that allows your code to perform multiple tasks simultaneously, thereby significantly enhancing its efficiency and speed.

Imagine you have a task at hand that's both time-consuming and computationally intensive. You're eager to optimize your Python code and make it run faster. That's where parallelism comes to the rescue. Parallelism is like having a team of skilled workers collaborating to accomplish a complex task much more quickly than a single worker could. In the world of software, it means breaking down a large task into smaller subtasks that can be executed simultaneously.

Python, known for its simplicity and versatility, offers multiple approaches to achieve parallelism. Whether you're dealing with CPU-bound tasks that involve intense computation or I/O-bound tasks that require waiting for external resources like network requests, Python has the tools and techniques to help you harness the power of parallelism.

One of the fundamental concepts of parallelism in Python is the use of threads. Threads are lightweight, independent units of execution within a process. They share the same memory space, making it easy for them to communicate and work together.

Python's **threading** module is an excellent starting point for parallelism. It allows you to create and manage threads effortlessly. Threads are well-suited for I/O-bound tasks where waiting for external resources like network requests or file operations is the primary bottleneck.

Consider a scenario where you have a web scraping task that involves fetching data from multiple websites. Traditionally, without parallelism, you'd fetch data from one website at a time, leading to inefficiencies. But with multithreading, you can fetch data from multiple websites concurrently, significantly speeding up the process.

Here's a basic Python script that demonstrates multithreading for fetching data from multiple websites:

pythonCopy code

```
import threading import requests # Define a list of websites
to fetch data from websites = ['https://example.com',
'https://python.org', 'https://github.com'] # Function to
fetch data from a website def fetch_website_data(url):
response = requests.get(url) print(f"Fetched data from {url},
Length: {len(response.text)}") # Create thread objects for
each website threads = [] for website in websites: thread =
threading.Thread(target=fetch_website_data,
args=(website,)) threads.append(thread) # Start the threads
for thread in threads: thread.start() # Wait for all threads to
finish for thread in threads: thread.join() print("All threads
have finished.")
```

In this script, we import the **threading** module and define a list of websites to fetch data from. We then create a function **fetch_website_data** that takes a URL as an argument and uses the **requests** library to fetch data from that URL. Next, we create thread objects for each website, passing the **fetch_website_data** function as the target.

By starting these threads and waiting for them to finish using the **start** and **join** methods, respectively, we enable concurrent fetching of data from multiple websites. This approach significantly reduces the time it takes to complete the task compared to fetching data sequentially.

However, it's crucial to understand that Python's Global Interpreter Lock (GIL) can limit the performance benefits of multithreading in CPU-bound tasks. The GIL restricts the execution of multiple threads in a single Python process, so in CPU-bound scenarios, multithreading might not provide the expected speedup. In such cases, multiprocessing or asynchronous programming may be more suitable.

Multiprocessing is another powerful way to achieve parallelism in Python, especially for CPU-bound tasks where computation is the bottleneck. Python's **multiprocessing** module allows you to create multiple processes, each with its own Python interpreter and memory space.

Let's consider an example where we calculate the factorial of multiple numbers concurrently using the **multiprocessing** module:

pythonCopy code

```
import multiprocessing # Function to calculate factorial of a number def calculate_factorial(num): result = 1 for i in range(1, num + 1): result *= i return result # List of numbers to calculate factorial for numbers = [5, 10, 15, 20] # Create a pool of processes pool = multiprocessing.Pool() # Map the function to the numbers and retrieve results results = pool.map(calculate_factorial, numbers) # Close the pool pool.close() pool.join() # Print
```

the results for num, result in zip(numbers, results): print(f"Factorial of {num} is {result}")

In this example, we define a function **calculate_factorial** that calculates the factorial of a given number. We then create a pool of processes using **multiprocessing.Pool** and use the **map** method to distribute the calculation of factorials for multiple numbers across these processes. This approach takes full advantage of multiple CPU cores and can provide significant speed improvements for CPU-bound computations.

Asynchronous programming is yet another powerful technique for achieving parallelism in Python, primarily for I/O-bound tasks that involve waiting for external resources, such as network requests or file operations. Python's **asyncio** library provides the tools to write asynchronous code using **async** and **await** keywords.

Here's a simple example of asynchronous web scraping using **asyncio** and the **aiohttp** library:

pythonCopy code

```
import asyncio import aiohttp # List of URLs to fetch data
from asynchronously urls = ['https://example.com',
'https://python.org', 'https://github.com'] async def
fetch_website_data(url): async with
aiohttp.ClientSession() as session: async with
session.get(url) as response: data = await response.text()
print(f"Fetched data from {url}, Length: {len(data)}") #
Create an event loop and gather tasks for fetching data
async def main(): tasks = [fetch_website_data(url) for url
in urls] await asyncio.gather(*tasks) # Run the event loop if
__name__ == '__main__': asyncio.run(main())
```

In this example, we define an asynchronous function **fetch_website_data** that fetches data from a URL using the **aiohttp** library. We create an event loop and gather tasks for fetching data from multiple URLs concurrently. The **asyncio.run(main())** line starts the event loop.

Asynchronous programming shines in scenarios where you have many I/O-bound operations, as it allows you to efficiently manage multiple tasks without the overhead of creating and managing threads or processes.

In summary, achieving parallelism in Python is a powerful way to enhance the performance of your applications, whether they involve CPU-bound computations or I/O-bound operations. Python offers a range of tools and libraries, including **threading, multiprocessing**, and **asyncio**, to help you harness the benefits of parallel programming.

By choosing the right approach and techniques for your specific task, you can unlock the potential of parallelism and make your Python code more responsive and efficient. So, take the plunge into the world of parallelism, and let Python be your ally in achieving parallelism for improved performance in your projects. Happy coding!

Chapter 9: Creating GUI Applications

Welcome to the captivating realm of graphical user interfaces (GUIs) in Python! In this chapter, we're about to embark on an exciting journey into the world of building intuitive and interactive interfaces for your Python applications. GUIs are the visual windows and elements that allow users to interact with your software effortlessly, making it more user-friendly and engaging.

Think about some of your favorite applications—whether it's a sleek, modern mobile app or a powerful desktop software. What sets them apart and makes them enjoyable to use? It's often the user interface—the buttons, menus, text fields, and graphics—that provides a seamless and visually appealing experience. Python, with its vast ecosystem of GUI libraries, allows you to create such interfaces, whether you're developing a desktop application, a web app, or even a game.

The need for graphical user interfaces arises when text-based interactions are no longer sufficient or practical. Imagine trying to control a complex video editing software or play a visually intensive game with just a command-line interface; it would be challenging and far from user-friendly. GUIs come to the rescue by providing a more natural and intuitive way to interact with software.

Python offers several libraries and frameworks for building GUIs, each with its own unique strengths and use cases. Some of the most popular options include Tkinter, PyQt, wxPython, and Kivy. Let's explore Tkinter, which is the standard GUI library included with Python and is well-suited for creating simple to moderately complex desktop applications.

Tkinter provides a rich set of widgets (UI elements) that you can use to design your application's interface. These widgets include buttons, labels, entry fields, text areas, and more. With Tkinter, you can arrange these widgets in windows and frames to create your application's layout.

Let's dive into a simple example to illustrate how to create a basic GUI application using Tkinter:

pythonCopy code

```
import tkinter as tk # Create a main application window
app = tk.Tk() app.title("My GUI App") # Create a label
widget label = tk.Label(app, text="Welcome to my GUI
app!") label.pack() # Create a button widget button =
tk.Button(app, text="Click me!") button.pack() # Define a
function to run when the button is clicked def
button_click(): label.config(text="Button clicked!") #
Configure the button to call the function when clicked
button.config(command=button_click) # Start the main
event loop app.mainloop()
```

In this example, we import the **tkinter** module, create a main application window with the **Tk()** constructor, and set its title. We then create a label and a button widget, adding them to the main window with the **pack()** method.

To make the button interactive, we define a function **button_click** that changes the label's text when the button is clicked. We configure the button to call this function when clicked using the **config** method. Finally, we start the main event loop with **app.mainloop()**, which keeps the GUI responsive and allows users to interact with it. This example provides a glimpse of how easy it is to create a simple GUI application with Tkinter. You can extend this by adding more widgets, arranging them in different layouts, and defining more complex interactions to suit your application's requirements. While Tkinter is a fantastic choice for desktop applications, Python also offers solutions for web-based GUIs. If you're interested in developing web applications with a Python backend and a user-friendly interface, libraries like Flask and Django can help. These web frameworks allow you to create web pages with HTML, CSS, and JavaScript, while Python handles the backend logic. Another exciting avenue in GUI development is game development. Python provides libraries like Pygame and Panda3D that enable you to build interactive 2D and 3D games with graphical interfaces. Whether you want to create a simple puzzle game or a sophisticated 3D adventure, Python has the tools to make it happen. In addition to desktop applications, web interfaces, and games, Python's GUI capabilities extend to mobile app development. Kivy, for example, is a versatile framework that lets you build cross-platform mobile apps with Python. Whether you're targeting iOS, Android, or both, Kivy provides the tools and widgets you need to create responsive and visually appealing mobile applications. No matter which path you choose—desktop applications, web interfaces, games, or mobile apps—Python's GUI libraries and frameworks offer the versatility and power to bring your software ideas to life. The key is to select the right tool for your project based on its requirements and target platform. Before we wrap up, let's not forget the importance of user experience (UX) and design in GUI development. A visually pleasing and intuitive interface can make all the difference in how users perceive and interact with your software. Consider factors like layout, color schemes, typography, and user workflows when designing your GUI. Tools like Adobe XD, Figma, or even simple sketches on paper can help you plan and visualize your interface before you start coding.

In summary, building graphical user interfaces in Python opens up a world of possibilities for creating user-friendly, interactive software. Whether you're developing desktop applications, web interfaces, games, or mobile apps, Python provides a wealth of libraries and frameworks to help you bring your ideas to fruition.

So, embrace the art of GUI development in Python, and let your creativity shine as you craft visually appealing and user-friendly interfaces for your projects. With the right tools and a dash of design sensibility, you can create software that not only works well but also delights users with its ease of use and aesthetics. Happy coding, and may your GUIs be both functional and beautiful! Welcome to the fascinating world of GUI (Graphical User Interface) design patterns and best practices! In this chapter, we'll explore the art and science of creating visually appealing, user-friendly interfaces that enhance the overall user experience of your applications. GUI design is a delicate balance between aesthetics and functionality, and it plays a pivotal role in how users perceive and interact with your software. Design patterns are recurring solutions to common design problems. In the context of GUI design, they serve as templates and guidelines for creating interfaces that are both intuitive and efficient. These patterns have evolved over time through the collective experience of designers and developers, and they provide a solid foundation for building interfaces that work seamlessly. One fundamental design principle that underlies most GUI design patterns is consistency. Consistency ensures that elements and interactions within your interface follow a predictable and uniform pattern. When users can anticipate how things will work based on their previous interactions, it reduces confusion and frustration.

A prime example of a GUI design pattern is the "Menu Bar." You've likely encountered this pattern in many desktop applications, where a horizontal bar at the top of the window contains menus such as "File," "Edit," and "View."

Clicking on these menus reveals a dropdown list of options. The Menu Bar provides a consistent and familiar way for users to access various functionalities of the application.

Another prevalent design pattern is the "Toolbar." Toolbars typically reside just below the Menu Bar and contain icons or buttons representing frequently used actions. Toolbars offer quick access to essential features, saving users the trouble of navigating through menus. These icons are often accompanied by tooltips to provide additional context when users hover over them.

Dialog boxes are a versatile design pattern used for various purposes, such as collecting user input, displaying notifications, or confirming actions. The "Modal Dialog Box" pattern is particularly useful when you need to temporarily interrupt the main workflow to address a specific task or decision. Modal dialogs force users to interact with them before returning to the main application, making them ideal for critical actions like saving changes or confirming deletions. Tabs and tabbed interfaces are another design pattern that helps organize content and keep the interface tidy. Tabs allow users to switch between different sections or views of an application without cluttering the screen. Each tab represents a distinct context or task, offering users a clear mental model of the interface's structure.

Now, let's talk about "Wizard" patterns. Wizards guide users through a series of steps or decisions, simplifying complex processes by breaking them into manageable chunks. Wizards often feature a "Next" button that leads users forward while providing clear instructions and feedback along the way. This pattern is valuable for tasks like software installation, setting up profiles, or creating documents with multiple configuration options.

The "Master-Detail" pattern is prevalent in applications that deal with hierarchical data or lists. In this pattern, the left side typically displays a list or overview (the "Master"), and the right side shows detailed information or actions related to the selected item (the "Detail"). This pattern helps users navigate and manipulate data efficiently, especially in applications like email clients, file managers, and contact lists.

Progress indicators and feedback elements are essential for keeping users informed about ongoing processes. The "Spinner" pattern, often represented by a spinning wheel or animated icon, signifies that the application is working on a task and prevents users from feeling stuck or uncertain during loading times or lengthy operations.

Whitespace, alignment, and visual hierarchy are vital concepts in GUI design. Whitespace, or empty space around and between elements, helps create a sense of balance and prevents overcrowding. Proper alignment ensures that elements are visually connected, making it easier for users to scan and understand the layout. Visual hierarchy, achieved through variations in size, color, and typography, guides users' attention to the most critical elements or actions.

Designing for accessibility is a critical aspect of GUI design. Accessibility ensures that your interface is usable by individuals with disabilities, such as visual or motor impairments. Common practices include providing alt text for images, ensuring keyboard navigation, using sufficient color contrast, and offering resizable text. Accessibility not only benefits users with disabilities but also enhances the overall usability of your application.

User testing and usability testing are essential steps in the design process. These tests involve real users interacting with your interface and providing feedback. Usability testing helps you identify areas where users may struggle, encounter confusion, or experience frustration. By observing users' actions and listening to their feedback, you can refine your design to better align with their needs and expectations.

Responsive design is crucial in today's multi-device world. Your GUI should adapt gracefully to various screen sizes, from large desktop monitors to mobile devices. Responsive design involves flexible layouts, scalable graphics, and thoughtful touch-friendly interactions. Ensuring a consistent user experience across different devices enhances usability and user satisfaction.

While GUI design patterns and best practices provide valuable guidance, it's essential to remember that each application is unique, and context matters. Tailor your design choices to the specific needs and goals of your project. The goal is to strike a balance between adhering to established patterns and allowing for creative solutions that cater to your users' particular requirements.

Incorporating user feedback and conducting usability tests throughout the design and development process can help you fine-tune your interface and address any issues or concerns. Remember that design is an iterative process, and continuous improvement is key to creating exceptional user experiences.

In summary, GUI design patterns and best practices serve as valuable tools in crafting user-friendly and visually appealing interfaces. They provide a foundation for creating interfaces that are not only aesthetically pleasing but also efficient and intuitive to use. By applying these patterns and principles while keeping the unique context of your project in mind, you can design interfaces that enhance the overall user experience and leave a positive impression on your users. So, let your creativity flourish, and may your GUIs be both functional and delightful!

Chapter 10: Intermediate Automation Projects and Case Studies

Welcome to the fascinating world of real-world automation examples, where we'll explore how Python can be a powerful ally in streamlining everyday tasks and solving practical problems. Automation is all about simplifying our lives and increasing productivity, and Python provides the perfect toolkit to achieve these goals.

Imagine this scenario: You receive a flood of emails daily, and buried in those emails are important attachments that need to be saved to a specific folder. Doing this manually would be tedious and time-consuming. But with Python, you can create a script that automatically scans your email, identifies attachments, and saves them to the designated location. This is just one example of how Python can simplify your daily routine. Another common task is data extraction. Let's say you regularly need to collect information from various websites or online sources. Manually copying and pasting this data into a spreadsheet can be a daunting task. Python's web scraping capabilities, combined with libraries like Beautiful Soup and Requests, can automate this process for you. You can write a script that navigates to web pages, extracts the data you need, and stores it in a structured format, ready for analysis. Have you ever faced the challenge of managing large sets of files and folders? Python can help you tidy up your file system with ease. You can write a script that organizes files based on criteria like file type, creation date, or keywords in their names. This can be a huge time-saver and prevent the frustration of searching for misplaced files. Python also excels at automating repetitive data manipulation tasks. Whether you're dealing with spreadsheets, databases, or text files, Python can quickly and accurately process data. For example, you might have a CSV file with thousands of records that require cleaning and formatting. Instead of doing this manually, you can write a Python script to perform tasks like removing duplicates, correcting data formats, and generating summary reports automatically.

Automation extends to the world of social media and marketing as well. Let's say you run a business and want to schedule posts on your social media accounts. Python has libraries like Tweepy for Twitter and Instabot for Instagram that allow you to automate posting at specific times or dates. This ensures that your content reaches your audience when it's most effective.

One of the more complex but rewarding automation tasks is building chatbots. Whether you want to provide customer support or automate responses to common queries, Python libraries like ChatterBot make it feasible. You can train your chatbot to understand and respond to user inputs, creating a more interactive and engaging experience for your website or application users.

Python is also a valuable tool for automating data analysis and reporting. For instance, you might have data coming in from various sources that need to be combined, cleaned, and analyzed regularly. With Python's data manipulation libraries like Pandas and visualization libraries like Matplotlib, you can create scripts that automate the entire process, from data collection to generating reports and visualizations.

For professionals in the finance industry, Python is indispensable for tasks like algorithmic trading and portfolio management. Libraries like QuantLib and Zipline enable you to automate trading strategies, backtest them with historical data, and make data-driven investment decisions.

Quality assurance and software testing benefit greatly from automation. Python's Selenium library allows you to write scripts for automating web application testing. You can simulate user interactions, verify that web pages behave correctly, and detect issues early in the development process.

Python's automation capabilities extend to the realm of system administration. You can create scripts to manage server tasks, monitor system performance, and perform routine maintenance. This not only saves time but also reduces the risk of human error in critical system operations.

In the world of cybersecurity, Python is a versatile tool for automating tasks like penetration testing, vulnerability scanning, and network monitoring. Security professionals use Python to identify vulnerabilities, assess security risks, and respond to threats in real-time.

Python's automation potential even extends to creative fields like art and music. You can write scripts to generate digital art, create music compositions, or even design 3D models. Artists and musicians use Python to explore new creative avenues and automate repetitive design or composition tasks.

In the Internet of Things (IoT) domain, Python plays a crucial role in automating the interactions between connected devices. You can use Python to collect data from sensors, control smart appliances, and build home automation systems that enhance convenience and energy efficiency.

Automation is not limited to standalone scripts. Python integrates seamlessly with existing software and tools through APIs (Application Programming Interfaces). This means you can automate interactions with popular software applications, such as Microsoft Office Suite, Google Workspace, and cloud services like AWS and Azure.

In summary, Python's versatility and extensive library ecosystem make it a powerhouse for real-world automation. Whether you're streamlining email management, extracting data from the web, organizing files, cleaning and analyzing data, or automating tasks in various domains like social media, finance, quality assurance, system administration, cybersecurity, creative arts, IoT, or software integration, Python provides the tools you need to simplify your life and boost productivity. Automation is not only about saving time but also about reducing errors, increasing efficiency, and unlocking new possibilities. As you embark on your automation journey with Python, remember that the only limit is your creativity. With Python as your automation companion, you have the potential to transform how you work, opening doors to more innovation and free time to focus on what truly matters to you. Happy automating!

Welcome to the world of intermediate-level automation, where we delve into real-world case studies that showcase the power and versatility of Python. These case studies serve as practical examples of how Python can be applied to solve complex problems, streamline workflows, and improve efficiency in various domains. Let's start with a case study in data analysis. Imagine you work for a retail company with a vast amount of sales data. Your task is to extract insights from this data to make informed business decisions. Python's data manipulation and visualization libraries, such as Pandas and Matplotlib, come to the rescue. You can write Python scripts to clean, analyze, and visualize the sales data, uncovering trends, identifying top-performing products, and optimizing pricing strategies. This data-driven approach helps your company increase sales and profitability.

Moving on to the world of finance, consider a scenario where you're an investment analyst responsible for managing a portfolio of stocks and bonds. Python's libraries for financial analysis, like QuantLib and NumPy, enable you to create scripts that automate portfolio optimization. These scripts can calculate the optimal asset allocation, taking into account risk and return objectives. This not only saves time but also leads to more informed investment decisions and better returns for your clients. In the field of healthcare, Python can be a valuable tool for medical research and analysis. Let's say you're working on a project to analyze medical images, such as X-rays or MRIs, to detect and diagnose diseases. Python's image processing libraries, like OpenCV and scikit-image, empower you to build image analysis pipelines. These pipelines can automatically detect abnormalities, segment organs, and provide quantitative data for research purposes. Your work contributes to early disease detection and better patient outcomes. For educators and e-learning platforms, Python can automate the creation of personalized learning experiences. Suppose you're designing an adaptive learning system that tailors content to each student's proficiency level. Python's machine learning libraries, such as scikit-learn, allow you to develop algorithms that analyze student performance and recommend appropriate lessons and exercises. This personalized approach enhances learning outcomes and engagement. In the realm of marketing, consider a situation where you manage a large customer database for a retail e-commerce platform. You want to create targeted email marketing campaigns based on customer preferences and behavior. Python's data analysis and machine learning capabilities enable you to segment customers into groups with similar characteristics. By automating the process of campaign selection and customization, you can increase conversion rates and customer satisfaction.

Python's automation prowess extends to the world of content generation. Imagine you're a content creator responsible for generating blog posts or articles on a regular basis. Python's natural language processing (NLP) libraries, such as NLTK and spaCy, can assist in automating content generation. You can build scripts that analyze trending topics, extract key insights from data, and generate high-quality written content. This not only saves time but also ensures that your content remains relevant and engaging.

In the logistics and supply chain industry, Python can optimize route planning and resource allocation. Suppose you work for a delivery company tasked with delivering packages to various locations. Python's optimization libraries, like PuLP and Gurobi, allow you to create scripts that determine the most efficient delivery routes while considering factors like traffic, package size, and delivery windows. This optimization leads to cost savings and improved delivery times.

Python is a valuable ally in quality assurance and software testing. Let's say you're part of a software development team responsible for testing a complex web application. Python's Selenium library enables you to automate web testing by simulating user interactions, verifying functionality, and generating detailed test reports. Automated testing reduces the risk of human errors and ensures the reliability of the software.

For environmental scientists and researchers, Python aids in data collection and analysis. Suppose you're conducting a study on air quality and need to collect data from multiple sensors placed throughout a city. Python scripts can automate data retrieval, aggregation, and visualization. Real-time monitoring and analysis of air quality data help identify pollution sources and guide environmental policies.

Python's role in game development is not to be overlooked. Game developers can automate various aspects of game design and testing. For example, Python scripts can generate random game levels, test gameplay mechanics, and analyze player behavior data. This automation accelerates the game development process and enhances game quality.

In the legal profession, Python can assist in document review and analysis. Consider a scenario where you need to review thousands of legal documents for a case. Python's text processing capabilities allow you to create scripts that automatically extract key information, identify relevant sections, and summarize the content. This automation saves countless hours and ensures thorough document analysis.

In the field of scientific research, Python is a versatile tool for conducting experiments and simulations. Whether you're studying physics, chemistry, biology, or any other scientific discipline, Python's scientific computing libraries, such as SciPy and SymPy, provide the computational power needed for complex simulations and data analysis.

These case studies illustrate the breadth and depth of Python's automation capabilities. From data analysis and finance to healthcare, education, marketing, content generation, logistics, quality assurance, environmental research, game development, legal analysis, and scientific research, Python is a reliable companion that empowers professionals to automate tasks, make data-driven decisions, and achieve remarkable results.

As you explore these case studies, remember that Python is not just a programming language; it's a versatile tool that can transform the way you work and open doors to innovative solutions. With Python as your automation partner, you have the potential to excel in your domain, improve efficiency, and make a positive impact in your field. So, let's embark on this automation journey together and discover how Python can elevate your work to new heights.

BOOK 3
PYTHON AUTOMATION MASTERY
ADVANCED STRATEGIES

ROB BOTWRIGHT

Chapter 1: Mastering Advanced Data Analysis with Python

Welcome to the realm of advanced data manipulation techniques, where we'll dive deep into the world of Python and explore how it can empower you to master the art of working with data like a seasoned data scientist. In our journey, we'll uncover advanced methods and strategies that go beyond the basics, allowing you to extract valuable insights from complex datasets and transform them into actionable knowledge.

One of the first techniques we'll explore is the art of reshaping data. Imagine you have a dataset with information stored in a wide format, where each column represents a variable, and you want to convert it into a long format. Python's Pandas library offers powerful tools like the **melt** function to reshape data efficiently. This reshaping process is crucial when dealing with datasets from different sources that require harmonization.

Now, let's delve into data aggregation. In real-world scenarios, you often deal with large datasets containing granular information, but what you need is a higher-level summary. Python's Pandas and NumPy libraries offer advanced aggregation functions like **groupby**, **pivot_table**, and **agg** to help you summarize data based on specific criteria. This is particularly useful for generating meaningful reports and visualizations.

Advanced data transformation techniques are essential for preparing data for analysis. Python provides tools for handling missing data, outliers, and anomalies. You can use techniques like interpolation, imputation, and robust statistical methods to ensure your data is clean and ready for analysis. This level of data preparation is critical for obtaining accurate insights.

Data integration is a common challenge in data science, especially when dealing with multiple datasets from diverse sources. Python's Pandas library offers powerful techniques for merging and joining datasets. Whether you need to perform inner joins, outer joins, or complex merging operations, Python has you covered. This allows you to combine data from different sources and derive comprehensive insights.

Advanced filtering and selection techniques are essential when working with extensive datasets. Python's Pandas library provides robust filtering capabilities, allowing you to extract specific rows and columns that meet complex criteria. You can also apply conditional logic to select data, making it easier to focus on relevant information.

Data sampling is a critical technique when you're dealing with massive datasets and want to work with a manageable subset. Python provides various sampling methods, such as random sampling, stratified sampling, and systematic sampling. These techniques allow you to extract representative samples for analysis, reducing computational load and processing time.

Time series analysis is a specialized area within data manipulation that deals with temporal data. Python's libraries, including Pandas, NumPy, and Statsmodels, offer advanced time series manipulation and analysis capabilities. You can resample time series data, calculate rolling statistics, and perform forecasting using state-of-the-art techniques.

Handling categorical data effectively is crucial in many data analysis tasks. Python's Pandas library provides advanced techniques for encoding categorical variables, such as one-hot encoding, label encoding, and target encoding. These methods enable you to represent categorical data in a format suitable for machine learning algorithms.

Feature engineering is an art in data science, and Python offers a wealth of tools and techniques to create new features from existing data. Whether you're generating interaction terms, polynomial features, or custom transformations, Python's libraries make it easy to engineer features that improve the performance of machine learning models.

Data transformation pipelines are a powerful way to automate and streamline data preprocessing tasks. Python's scikit-learn library provides a robust framework for building data transformation pipelines. These pipelines allow you to define a sequence of data preprocessing steps and apply them consistently to new data, ensuring consistency and reproducibility.

Advanced data visualization techniques can help you gain deeper insights from your data. Python's libraries, such as Matplotlib, Seaborn, and Plotly, offer a wide range of visualization options. You can create interactive plots, heatmaps, 3D visualizations, and more to explore data from different angles and uncover hidden patterns.

Dealing with unstructured data, such as text or images, requires specialized techniques. Python's libraries, including NLTK, spaCy, and scikit-image, offer advanced text and image processing capabilities. You can perform natural language processing tasks like text classification, sentiment analysis, and topic modeling, as well as image recognition and segmentation.

Advanced statistical analysis techniques, such as hypothesis testing, regression analysis, and machine learning, are at your fingertips with Python. Libraries like Statsmodels, scikit-learn, and TensorFlow provide the tools you need to perform sophisticated statistical analysis and build predictive models.

These techniques allow you to extract valuable insights and make data-driven decisions.

Parallel processing and distributed computing can significantly speed up data manipulation tasks, especially when dealing with large datasets. Python's libraries, like Dask and multiprocessing, enable you to leverage multiple CPU cores and distributed computing environments. This enhances the scalability and efficiency of data manipulation operations.

Data version control is essential for tracking changes to datasets and ensuring reproducibility. Python's libraries, such as GitPython and DVC (Data Version Control), provide tools for managing data versioning and collaboration. This ensures that you can track and reproduce data manipulation steps, making your work more transparent and reliable.

Incorporating data manipulation into a data science workflow is a critical skill. Python offers integration with popular data science platforms like Jupyter notebooks and cloud-based environments like Google Colab. This allows you to seamlessly integrate data manipulation into your data analysis and modeling processes.

In summary, advanced data manipulation techniques in Python are the key to unlocking the full potential of your data. Whether you're reshaping data, aggregating information, integrating datasets, performing advanced filtering, or handling time series and categorical data, Python provides a rich toolbox of libraries and techniques. These tools empower you to transform raw data into valuable insights, making informed decisions and driving innovation in your field. So, let's roll up our sleeves and dive deeper into the world of advanced data manipulation with Python! Welcome to the fascinating world of statistical analysis and modeling with Python. In this chapter, we will embark on a journey through the realm of data-driven decision-making, where Python serves as our trusty guide. Along the way, we will explore how statistical techniques and modeling can transform raw data into meaningful insights, empowering you to make informed choices in various domains.

Statistical analysis is the foundation of data science and plays a pivotal role in understanding patterns, relationships, and trends in data. Imagine you have a dataset containing information about customer preferences for an e-commerce platform. You can employ Python's statistical libraries, such as NumPy and SciPy, to calculate descriptive statistics like mean, median, and standard deviation. These summary statistics offer a snapshot of your data's central tendencies and dispersion.

When dealing with uncertainty and variability, hypothesis testing becomes a valuable tool. Python's Statsmodels library provides a robust framework for conducting hypothesis tests, such as t-tests, chi-squared tests, and ANOVA. These tests allow you to determine whether observed differences or relationships in your data are statistically significant. Whether you're comparing sales figures before and after a marketing campaign or evaluating the effectiveness of a new drug, hypothesis testing helps you draw reliable conclusions.

Regression analysis is another cornerstone of statistical modeling. Python's Statsmodels and scikit-learn libraries offer a wide range of regression techniques, including linear regression, logistic regression, and polynomial regression. Regression models help you uncover relationships between variables, make predictions, and understand how changes in one variable affect others. For instance, you can build a linear regression model to predict a house's price based on its square footage, number of bedrooms, and other features.

Time series analysis is essential when working with temporal data, such as stock prices, weather measurements, or economic indicators. Python's libraries, including Pandas and Statsmodels, provide specialized tools for time series modeling. You can perform tasks like decomposition, autocorrelation analysis, and forecasting. Time series models help you uncover patterns in data that evolve over time and make predictions for future values.

Moving beyond traditional statistical techniques, machine learning models offer a powerful way to make predictions and classifications. Python's scikit-learn and TensorFlow libraries provide a vast array of machine learning algorithms, from decision trees and support vector machines to neural networks and deep learning models. These algorithms allow you to build predictive models for tasks like image recognition, sentiment analysis, fraud detection, and more. Machine learning empowers you to extract valuable insights from data and automate decision-making processes.

Clustering and dimensionality reduction techniques are invaluable for exploring data with many variables or identifying hidden patterns in large datasets. Python's scikit-learn offers algorithms like K-means clustering and principal component analysis (PCA) to group similar data points and reduce the dimensionality of data while preserving its essential characteristics. These techniques are crucial when dealing with high-dimensional data or trying to segment customers based on their behavior.

Bayesian statistics introduces a probabilistic approach to modeling uncertainty and updating beliefs as new data becomes available. Python's PyMC3 library allows you to perform Bayesian modeling, including Bayesian regression, Bayesian network analysis, and Markov chain Monte Carlo (MCMC) simulations. Bayesian methods are particularly useful in scenarios where you want to incorporate prior knowledge or continuously update models with incoming data, such as predicting disease outbreaks or estimating financial risks.

Survival analysis is a specialized statistical technique used to analyze time-to-event data, such as the time until a customer makes a purchase or the survival time of patients in a clinical trial. Python's lifelines library provides tools for survival analysis, allowing you to estimate survival curves, hazard rates, and perform hypothesis tests on survival data. Survival analysis is widely applied in healthcare, finance, and reliability engineering.

Anomaly detection is crucial for identifying unusual or unexpected events in data. Python's libraries, including scikit-learn and PyOD, offer algorithms for anomaly detection, such as isolation forests and one-class SVMs. Anomaly detection is essential for fraud detection, network security, and quality control, where identifying outliers can be of critical importance.

Ensemble methods combine multiple models to improve predictive accuracy and robustness. Python's scikit-learn library provides ensemble techniques like random forests, gradient boosting, and bagging. Ensemble methods are valuable when you want to reduce overfitting and improve the generalization performance of your models. They are widely used in machine learning competitions and real-world applications.

Model evaluation and validation are essential steps in the modeling process. Python's scikit-learn offers tools for cross-validation, hyperparameter tuning, and model selection. These techniques help you assess the performance of your models, choose the best hyperparameters, and avoid overfitting. Proper model evaluation ensures that your predictions are reliable and can be trusted in decision-making.

Time series forecasting is a specialized application of statistical modeling. Python's Prophet library, developed by Facebook, simplifies time series forecasting by providing an intuitive interface and handling holidays and seasonality automatically. Whether you're forecasting stock prices, demand for products, or website traffic, Prophet can help you generate accurate and interpretable forecasts.

Text analysis and natural language processing (NLP) techniques enable you to extract insights from unstructured text data, such as customer reviews, social media posts, or news articles. Python's NLP libraries, like NLTK, spaCy, and gensim, offer tools for text preprocessing, sentiment analysis, topic modeling, and named entity recognition. Text analysis is essential for understanding customer sentiment, identifying emerging topics, and automating text-based tasks. Geospatial analysis allows you to work with geographical data and perform tasks like geocoding, spatial visualization, and spatial analysis. Python's geospatial libraries, including Geopandas and Folium, provide tools for handling geospatial data formats, creating interactive maps, and conducting spatial analysis. Geospatial analysis is valuable in fields such as urban planning, environmental monitoring, and location-based services.

In summary, statistical analysis and modeling in Python offer a rich toolkit for extracting insights from data, making predictions, and understanding complex relationships. Whether you're performing hypothesis tests, building regression models, leveraging machine learning algorithms, or exploring time series data, Python provides the flexibility and power you need to tackle a wide range of data-driven challenges. So, let's dive deeper into the world of statistical analysis and modeling, where data comes to life, and decisions are made with confidence.

Chapter 2: Machine Learning and Data Science Foundations

Welcome to the exciting world of machine learning! In this chapter, we'll embark on a journey through the fascinating field of artificial intelligence and explore how machines can learn from data to make predictions and decisions. It's a journey that will take us from the fundamentals of machine learning to the cutting-edge advancements in the field, and along the way, we'll demystify the concepts and techniques that power this technology.

Machine learning is at the heart of the technological revolution of our era. It's the driving force behind recommendation systems, virtual personal assistants, autonomous vehicles, and so much more. But what exactly is machine learning, and how does it work? At its core, machine learning is about creating algorithms that can automatically learn patterns and make predictions or decisions based on data. These algorithms are designed to improve their performance as they are exposed to more data, and that's what makes them "learn."

To understand machine learning better, let's start with the two main types of machine learning: supervised learning and unsupervised learning. In supervised learning, the algorithm is provided with a labeled dataset, which means that each data point is associated with a target or outcome. The goal of the algorithm is to learn a mapping from the input data to the target labels, allowing it to make predictions on new, unseen data. This is commonly used in tasks like image classification, spam email detection, and predicting house prices.

Unsupervised learning, on the other hand, deals with unlabeled data. The algorithm's objective is to find patterns, structure, or relationships within the data without any predefined labels. Clustering and dimensionality reduction are common tasks in unsupervised learning. For instance, you could use unsupervised learning to group similar customers together based on their purchase behavior or to reduce the number of features in a dataset while preserving its essential information.

Now, let's talk about one of the most fundamental concepts in machine learning: the model. A model is a mathematical representation of a problem or a system that can make predictions or decisions. Models come in various forms, from simple linear regression models to complex deep neural networks. The choice of model depends on the problem you're trying to solve and the characteristics of your data.

To train a machine learning model, you need data—lots of it. The more data you have, the better your model can learn patterns and generalize to new, unseen data. However, data quality is just as important as quantity. Clean, well-structured data is the foundation of successful machine learning projects. It's like the raw material from which your model will learn.

In machine learning, we often divide our dataset into two parts: the training set and the test set. The training set is used to train the model, while the test set is kept separate for evaluating its performance. This separation allows us to assess how well the model will generalize to new, unseen data. It's a crucial step in ensuring that your model is not just memorizing the training data but actually learning useful patterns.

Now, let's talk about the process of training a machine learning model. During training, the algorithm adjusts its internal parameters to minimize a specific objective function, such as the mean squared error for regression tasks or the cross-entropy loss for classification tasks. This optimization process is often performed using techniques like gradient descent, which iteratively updates the model's parameters to find the best fit to the training data.

But how do you know if your model is performing well? That's where evaluation metrics come in. For regression tasks, you might use metrics like mean absolute error or R-squared to measure how well your model's predictions match the actual values. For classification tasks, metrics like accuracy, precision, recall, and F1 score help you assess the model's ability to correctly classify data points.

Model performance isn't static; it can vary depending on the specific problem and dataset. Hyperparameter tuning is the process of finding the best configuration for your model, such as the learning rate, the number of hidden layers in a neural network, or the choice of a kernel in a support vector machine. Techniques like grid search and random search help you explore different hyperparameter settings to optimize your model's performance.

Feature engineering is another critical aspect of machine learning. It involves selecting and transforming the input features that your model uses to make predictions. Good feature engineering can have a significant impact on your model's performance. For example, in a natural language processing task, you might extract features like word counts or word embeddings from text data.

Ensemble methods are a powerful technique in machine learning. They involve combining multiple models to improve overall performance. One common ensemble method is the random forest, which consists of multiple decision trees. By aggregating the predictions of these trees, random forests can reduce overfitting and increase the accuracy of predictions. Ensemble methods are widely used in competitions and real-world applications.

Deep learning, a subset of machine learning, has gained immense popularity in recent years. Deep neural networks, inspired by the human brain's structure, have achieved remarkable results in tasks like image recognition and natural language processing. Deep learning models, with their multiple layers of interconnected neurons, can automatically learn hierarchical features from data. However, they require large amounts of data and computational resources.

Transfer learning is a technique in deep learning that allows you to leverage pre-trained models on new, related tasks. Instead of starting from scratch, you can use a pre-trained model as a starting point and fine-tune it for your specific problem. Transfer learning has made it easier for researchers and practitioners to apply deep learning to a wide range of applications.

Reinforcement learning is a type of machine learning that focuses on training agents to make decisions by interacting with an environment. It's often used in tasks like game playing and robotics. Reinforcement learning agents learn to take actions that maximize a reward signal, which can lead to complex behaviors and strategies.

Machine learning isn't just about building models; it's also about deploying them in real-world applications. Model deployment involves integrating your trained model into a production environment where it can make predictions or decisions in real-time. This often requires considerations like scalability, latency, and monitoring.

In this chapter, we've covered the fundamental concepts of machine learning, from supervised and unsupervised learning to training models, evaluating performance, and deploying them in real-world scenarios. Machine learning is a vast and exciting field with countless applications, and we're just scratching the surface. As we dive deeper into the world of machine learning, we'll explore specific algorithms, techniques, and use cases that will empower you to tackle real-world challenges and harness the power of data-driven decision-making. So, let's continue our journey into the heart of machine learning, where data transforms into knowledge and predictions shape the future.

Data preprocessing and feature engineering are essential steps in preparing your data for machine learning, and they play a crucial role in the success of your models. In this chapter, we'll dive into the world of data preparation, where we'll learn how to clean, transform, and engineer features to make your data more suitable for machine learning tasks.

Before you feed your data into a machine learning model, it's essential to ensure that it's clean and well-structured. Real-world data can be messy, containing missing values, outliers, and inconsistencies. Cleaning your data involves identifying and handling these issues to create a more reliable dataset.

One common issue in data is missing values. These can occur for various reasons, such as data collection errors or incomplete records. Dealing with missing values is critical because many machine learning algorithms cannot handle them. You can choose to remove rows with missing values or impute them with a specific value or a statistical measure like the mean or median.

Outliers are data points that deviate significantly from the rest of the data. They can skew the results of your machine learning model, so it's important to detect and handle them appropriately. You can use techniques like the Z-score or the interquartile range (IQR) to identify outliers and then decide whether to remove them or transform them to reduce their impact.

Data consistency is another aspect of data cleaning. It involves checking for inconsistencies in your data, such as conflicting information or data that violates certain constraints. Addressing data consistency issues may require domain knowledge or additional data validation checks.

Once your data is clean, the next step is feature engineering. Feature engineering is the process of creating new features or transforming existing ones to make them more informative for your machine learning model. Well-engineered features can significantly improve your model's performance.

Feature engineering often starts with domain knowledge. You need to understand the problem you're solving and the data you're working with to identify relevant features. For example, in a predictive maintenance task for manufacturing, you might engineer features related to machine usage patterns, maintenance history, and sensor readings.

Feature scaling is a common preprocessing step. It ensures that all features have the same scale, preventing some features from dominating others in the model. Common scaling techniques include standardization (scaling to have zero mean and unit variance) and min-max scaling (scaling to a specific range, usually [0, 1]).

Categorical variables, which represent discrete categories or labels, require special treatment. Most machine learning algorithms work with numerical data, so you need to encode categorical variables into a numerical format. One-hot encoding is a popular method that converts each category into a binary vector, with a 1 indicating the presence of a category and 0 indicating its absence.

Another important aspect of feature engineering is dimensionality reduction. High-dimensional data can lead to increased computational complexity and overfitting. Principal Component Analysis (PCA) and t-Distributed

Stochastic Neighbor Embedding (t-SNE) are techniques commonly used for dimensionality reduction.

In some cases, you may want to create interaction features that capture relationships between different features. For example, in a housing price prediction task, you might create an interaction feature between the number of bedrooms and the number of bathrooms to capture the idea that larger houses with more bedrooms and bathrooms tend to have higher prices.

Feature selection is the process of choosing the most relevant features for your machine learning model. Removing irrelevant or redundant features can simplify your model, improve its performance, and reduce the risk of overfitting. Feature selection methods include statistical tests, feature importance scores, and recursive feature elimination.

Text data often requires specialized preprocessing. Textual features can be transformed into numerical representations using techniques like bag-of-words (BoW) or word embeddings (e.g., Word2Vec or GloVe). Text preprocessing also involves tasks like tokenization, stop word removal, and stemming or lemmatization to prepare text data for analysis.

Handling time-series data is another challenge in feature engineering. Time-series features can include lagged values, moving averages, and seasonality components. Creating meaningful time-related features can improve the predictive power of your models in tasks like stock price forecasting or weather prediction.

Feature engineering is not a one-time process but an iterative one. You may need to revisit and refine your features as you gain more insights into your data or experiment with different models. The goal is to create a set of features that captures the essential information in your data while avoiding noise and redundancy.

In summary, data preprocessing and feature engineering are essential steps in preparing your data for machine learning. Cleaning and transforming your data, encoding categorical variables, and engineering relevant features can significantly impact your model's performance. These techniques require a combination of domain knowledge, creativity, and experimentation to achieve the best results. So, roll up your sleeves and get ready to dive into the exciting world of data preparation and feature engineering!

Chapter 3: Deep Dive into Data Visualization

Welcome to the world of advanced data visualization libraries! In this chapter, we'll explore some of the most powerful and versatile tools available for creating compelling data visualizations. Data visualization is an art form that transforms raw numbers and statistics into meaningful insights, making it an indispensable tool for data scientists, analysts, and decision-makers.

Matplotlib is often considered the go-to library for creating static, publication-quality visualizations in Python. With Matplotlib, you can create a wide range of charts and plots, including line plots, scatter plots, bar charts, histograms, and more. Its flexibility and customization options allow you to fine-tune every aspect of your visualizations, from colors and labels to axis scales and annotations.

One of Matplotlib's strengths is its object-oriented approach to creating plots. You can build complex visualizations by creating and arranging individual components like axes, figures, and subplots. This level of control is especially useful when you need precise control over the layout and appearance of your plots.

Seaborn is a high-level data visualization library built on top of Matplotlib. It simplifies the process of creating attractive statistical visualizations by providing a high-level interface and a wide range of built-in themes and color palettes. Seaborn is particularly well-suited for tasks like creating distribution plots, heatmaps, and pair plots for exploring relationships between variables.

Pandas is another essential library for data manipulation and analysis in Python, but it also offers some powerful visualization capabilities. You can easily create basic plots like line charts, bar plots, and histograms directly from

Pandas DataFrames and Series. This integration with data structures makes Pandas a convenient choice for quick exploratory data analysis.

Plotly is a versatile library that excels in creating interactive visualizations for web applications and dashboards. It supports a wide range of chart types, including scatter plots, line charts, bar charts, and 3D visualizations. What sets Plotly apart is its interactivity—you can zoom, pan, hover over data points for details, and even add custom interactivity to your plots.

Bokeh is another library designed for interactive data visualization, with a focus on creating web-based, interactive visualizations. Bokeh provides a Pythonic interface for generating dynamic visualizations that can be embedded into web applications. Its support for interactivity, linked plots, and widgets makes it a valuable tool for building interactive dashboards.

Altair is a declarative statistical visualization library that simplifies the creation of interactive visualizations. With Altair, you specify your visualizations using a concise and intuitive syntax, and the library generates the corresponding Vega-Lite or Vega specifications. This approach makes it easy to create complex visualizations with minimal code.

D3.js is a powerful JavaScript library for creating custom, interactive data visualizations on the web. While it's not a Python library per se, it's widely used in web development, and there are Python wrappers like PyD3 that allow you to harness the power of D3.js in a Python environment. D3.js gives you complete control over the look and behavior of your visualizations, making it suitable for creating highly customized data-driven web applications.

Plotnine is a Python implementation of the Grammar of Graphics, a system for creating visualizations by specifying their structure and components. If you're familiar with ggplot2 in R, Plotnine will feel like a familiar tool. It allows you to express complex visualizations concisely by defining mappings between data and visual elements.

Holoviews is a Python library that simplifies data visualization by allowing you to declare visualizations as Python objects. With Holoviews, you can create a wide range of visualizations, including scatter plots, heatmaps, and geographic maps, with concise and declarative code. Its compatibility with multiple plotting libraries, including Matplotlib and Bokeh, makes it a flexible choice.

Datawrapper is an online tool that simplifies the process of creating visually appealing charts and maps for publications and presentations. While it's not a Python library, it's worth mentioning because of its ease of use and ability to generate embeddable visualizations for web use.

Advanced data visualization libraries are invaluable tools for transforming data into actionable insights. Whether you're creating static, publication-ready charts or building interactive web-based dashboards, there's a library that suits your needs. The choice of library often depends on factors like the complexity of your data, the level of interactivity required, and your familiarity with the library's syntax and capabilities.

In this chapter, we've introduced you to a range of advanced data visualization libraries, each with its strengths and use cases. As you delve further into data visualization, you'll discover that combining different libraries and tools can lead to even more powerful and expressive visualizations. So, let your data come to life, tell its story, and guide your decision-making with the help of these advanced data visualization tools. Happy visualizing!

Welcome to the captivating world of custom visualization and storytelling! In this chapter, we'll explore the art of crafting unique and compelling data visualizations that not only convey information but also tell a story. Data visualization is not just about creating charts and graphs; it's about communicating a narrative that resonates with your audience and drives meaningful insights.

Custom visualization goes beyond the standard chart types and templates offered by visualization libraries. It involves creating tailor-made visual representations that are designed to address specific questions or convey a particular message. Custom visualizations often require a deep understanding of the data and the audience's needs.

Storytelling is an integral part of effective data visualization. It's the process of structuring your visualizations in a way that guides the viewer through a narrative. A well-crafted data story has a clear beginning, middle, and end, taking the audience on a journey from data exploration to insights and conclusions.

To create custom visualizations, you need to start by understanding your data thoroughly. What are the key patterns, trends, and relationships hidden within the data? What story do you want to tell? Once you have a clear idea, you can choose the most appropriate visualization techniques to convey your message.

Custom visualizations often involve combining multiple chart types or modifying existing ones. For example, you might create a dual-axis chart that combines a line chart with a bar chart to show the relationship between two variables with different scales. This customization allows you to highlight correlations that might be obscured in a standard chart.

Another approach to custom visualization is creating animated or interactive visualizations. Animation can be a powerful storytelling tool, allowing you to illustrate changes over time or transitions between different states of the data. Interactive elements enable viewers to explore the data themselves, adding a layer of engagement to the visualization.

Infographics are a popular form of custom visualization that combines text, images, and visual elements to convey complex information in a concise and engaging manner. Infographics are often used for presenting data-driven stories in a format that is easy to digest and share.

Custom visualizations also involve attention to design principles. The choice of colors, fonts, and layout can significantly impact the effectiveness of your visualizations. Consistency in design elements across your visualizations helps create a cohesive and professional look.

Annotations and labels play a crucial role in custom visualization. They provide context and guide the viewer's attention to specific data points or insights. Well-placed annotations can help tell the story more effectively and clarify complex visualizations.

Data storytelling is about creating a narrative structure for your visualizations. Start with a clear and concise title that sets the stage for your story. Introduce the problem or question you're addressing, and explain why it matters. As you present your data, use captions and descriptions to provide context and interpretation.

Incorporating real-world examples and anecdotes can make your data story more relatable and memorable. Sharing stories of how data-driven insights led to actionable decisions or outcomes can inspire your audience and emphasize the value of data analysis.

Engaging your audience emotionally is a powerful storytelling technique. Use visuals, anecdotes, or personal stories to create an emotional connection to the data. Whether it's illustrating the impact of a social issue or celebrating a business success, emotions can make your story resonate more deeply.

Custom visualizations and storytelling are not limited to static images or presentations. They can be incorporated into data-driven reports, interactive dashboards, blog posts, or even video presentations. The choice of medium depends on your audience and the best way to convey your message.

Effective data visualization and storytelling require iteration and feedback. Don't be afraid to refine your visualizations and narratives based on audience feedback or new insights. Testing your visualizations with a sample audience can help you identify areas for improvement.

In summary, custom visualization and storytelling are essential skills for data professionals and communicators. They allow you to go beyond standard charts and graphs to create compelling narratives that resonate with your audience. Whether you're presenting data to inform decisions, advocate for change, or simply share insights, the art of custom visualization and storytelling can elevate your message and make your data come alive. So, embrace your creativity, explore new visualization techniques, and craft data stories that captivate and inspire!

Chapter 4: Web Application Development with Django

Welcome to the exciting world of web application development with Django! In this chapter, we'll embark on a journey to explore how Django, a high-level Python web framework, can empower you to build robust and feature-rich web applications. Whether you're a beginner or an experienced developer, Django provides a solid foundation for turning your web development dreams into reality.

Django is often hailed for its "batteries-included" philosophy, meaning it comes packed with a comprehensive set of tools, libraries, and features that streamline web development. This philosophy allows you to focus on building your application's unique features without getting bogged down in the nitty-gritty details of web development.

One of Django's core strengths is its built-in Object-Relational Mapping (ORM) system. This ORM abstracts the database layer, enabling you to work with your database using Python objects instead of writing raw SQL queries. This abstraction not only simplifies database interactions but also makes your code more portable across different database systems.Django's Model-View-Controller (MVC) architecture, often referred to as Model-View-Template (MVT) in Django terminology, promotes a clean separation of concerns. Models define the data structure, views handle the presentation logic, and templates are responsible for generating the HTML to be displayed. This separation makes your code more organized and maintainable. Another standout feature of Django is its built-in admin interface. With just a few lines of code, you can create a powerful administrative panel for managing your application's data. This admin interface is highly customizable, allowing you to tailor it to your specific needs and brand it with your application's style. Django's URL routing system is another gem. It allows you to define URL patterns that map to views, making it easy to create clean and human-readable URLs for your web application. This is not only user-friendly but also beneficial for search engine optimization (SEO).

Security is a top priority in web development, and Django takes it seriously. The framework includes built-in protection against common web vulnerabilities, such as Cross-Site Scripting (XSS), Cross-Site Request Forgery (CSRF), and SQL injection. This means you can focus on building your application's functionality while Django handles security at the framework level. Django's authentication system simplifies user management and authentication. You can quickly integrate user registration, login, and password reset functionalities into your web application. Additionally, Django's permission system allows you to control access to specific views or parts of your application with ease.

Django's internationalization and localization features make it accessible to a global audience. You can translate your application's content into multiple languages, making it more inclusive and user-friendly for a diverse user base.

When it comes to working with forms, Django shines. The framework provides a powerful and flexible form handling system that includes built-in form validation, field types, and customizable form rendering. Whether you need to create user registration forms or complex data entry forms, Django has you covered. Django's templating engine, known as the Django template language, makes it straightforward to generate dynamic HTML content. You can include variables, control structures, and even create custom template tags to extend its functionality. For those looking to work with APIs, Django offers the Django Rest Framework (DRF), a powerful toolkit for building Web APIs. DRF simplifies API development by providing serialization, authentication, and view classes out of the box, allowing you to focus on defining your API's endpoints and logic.

Django's extensibility is another remarkable feature. You can easily integrate third-party packages and libraries into your Django project, extending its functionality as needed. This extensibility allows you to leverage a vast ecosystem of open-source Django packages for tasks like user authentication, payment processing, and more.

Deployment is a critical aspect of web development, and Django provides guidance on deploying your applications to various hosting platforms, including traditional web servers and cloud providers. Django's documentation offers detailed instructions on how to configure and optimize your deployment for production use.

In summary, Django is a robust and versatile web framework that empowers developers to build web applications quickly and efficiently. Its rich set of features, strong focus on security, and emphasis on clean code architecture make it an ideal choice for a wide range of web projects, from simple websites to complex web applications. Whether you're just starting your web development journey or looking to level up your skills, Django is a friendly and powerful companion on your path to web application excellence. So, let's dive in and start building amazing web applications together!

Welcome to the exciting world of Django's advanced features and deployment strategies! In this chapter, we'll delve deeper into the Django framework, exploring its more advanced capabilities and discussing how to deploy your Django applications to make them accessible to the world.

As you become more proficient with Django, you'll discover a wealth of advanced features that can help you build even more powerful and dynamic web applications. One of these features is Django's support for complex database relationships, including many-to-many and one-to-one relationships. These advanced relationships enable you to model intricate data structures with ease, accommodating a wide range of application needs.

Django's middleware system is another advanced feature that provides a flexible way to process requests and responses globally in your application. By writing custom middleware, you can add functionalities like authentication checks, logging, or custom headers to every request-response cycle. This level of control allows you to fine-tune your application's behavior to meet specific requirements.

To take your application's user experience to the next level, Django offers support for handling asynchronous tasks through Django Channels. This feature is invaluable when you need real-time features like chat applications, notifications, or live updates. By integrating Django Channels, you can enable bidirectional communication between the server and the client in a scalable and efficient way.

Django's built-in support for caching is a powerful tool for optimizing your application's performance. By caching frequently used data or rendered templates, you can reduce the load on your database and speed up response times. Django supports various caching backends, including in-memory caches, file-based caches, and distributed caches like Redis.

When it comes to managing large datasets or complex queries, Django's database optimization techniques can make a significant difference in your application's performance. You can leverage features like database indexing, query optimization, and database-level constraints to ensure that your application remains responsive, even as your data grows.

Django's support for migrations simplifies database schema changes and versioning. With migrations, you can make changes to your database schema in a structured and automated way, ensuring that your database stays in sync with your application's evolving data model. This feature is particularly valuable in collaborative development environments where multiple developers may be working on the same project simultaneously.

As your Django project matures, you might find it necessary to implement a robust testing strategy. Django provides a comprehensive testing framework that allows you to write unit tests, integration tests, and even end-to-end tests for your application. By maintaining a suite of tests, you can ensure that new changes don't introduce regressions and that your application functions as expected.

Authentication and authorization are critical aspects of web application security. Django's advanced authentication options include support for social authentication via third-party providers like Google, Facebook, or GitHub. Additionally, you can implement custom authentication backends and fine-grained authorization policies to control user access to different parts of your application.

Django's built-in support for internationalization and localization goes beyond basic translation. You can easily create multilingual applications with translated content, date formats, and time zones. This advanced feature enables you to reach a global audience and provide a user-friendly experience in multiple languages.

Django's extensibility is a core principle, and as you advance in your Django journey, you'll likely explore the world of reusable Django apps. These apps, created by the Django community, cover a wide range of functionalities, from user authentication and content management to e-commerce and geospatial applications. Integrating these apps into your project can save you time and effort while adding valuable features.

Deployment is the final frontier in your Django adventure. Once you've built your application, you'll want to make it accessible to the world. Django offers various deployment options, from traditional web hosting using Apache or Nginx to modern cloud platforms like AWS, Heroku, or Google Cloud. The choice of deployment strategy depends on your application's scalability and performance requirements.

When deploying your Django application, you'll need to consider factors like database management, media storage, and server configuration. Docker and containerization can simplify deployment by packaging your application and its dependencies into a single container, ensuring consistent and reproducible deployments across different environments.

Monitoring and maintenance are ongoing tasks to keep your deployed application healthy and performant. Tools like Django Debug Toolbar and database query profilers can help you identify performance bottlenecks and troubleshoot issues in your production environment.

In summary, Django's advanced features and deployment options open up a world of possibilities for building robust and feature-rich web applications. As you continue to explore Django's capabilities, you'll gain the skills and knowledge needed to tackle complex application requirements and deliver exceptional user experiences. So, embrace the power of Django, experiment with its advanced features, and deploy your applications to the world with confidence. Your Django journey is an exciting adventure that leads to creating impressive web applications that make a difference!

Chapter 5: Cloud Integration and Deployment

Welcome to the fascinating world of cloud services and integration options! In this chapter, we'll dive into the realm of cloud computing and explore how it can enhance your applications, providing scalability, flexibility, and robustness that traditional on-premises solutions might struggle to match.

Cloud computing has revolutionized the way applications are built, deployed, and maintained. It offers a wide range of services and resources that can help you build, scale, and manage your applications more efficiently. Let's start by understanding what cloud computing is all about.

At its core, cloud computing is the delivery of computing services over the internet. These services can include computing power, storage, databases, networking, analytics, and more. Instead of owning and maintaining physical hardware, you can leverage cloud resources on a pay-as-you-go basis, which is often referred to as Infrastructure as a Service (IaaS).

One of the major advantages of cloud computing is scalability. Cloud providers offer the ability to scale your resources up or down based on demand, which means you can handle traffic spikes without worrying about hardware limitations. This elasticity is crucial for applications that experience fluctuating workloads.

Cloud providers, such as Amazon Web Services (AWS), Microsoft Azure, and Google Cloud Platform (GCP), offer a vast array of services that can be integrated into your applications. These services cover a wide range of use cases, from machine learning and artificial intelligence to data analytics, content delivery, and more.

Serverless computing is an exciting trend in cloud computing. It allows you to run code without managing servers, focusing solely on writing your application's logic. AWS Lambda, Google Cloud Functions, and Azure Functions are examples of serverless platforms that can execute your code in response to events or HTTP requests.

Containers and container orchestration tools like Docker and Kubernetes have become essential components of cloud-native development. Containers package your application and its dependencies into a single, lightweight unit that can run consistently across different environments. Kubernetes simplifies the deployment, scaling, and management of containerized applications.

Cloud providers also offer managed databases that eliminate the need for you to handle database maintenance tasks. You can choose from a variety of database options, including relational databases like MySQL and PostgreSQL, as well as NoSQL databases like MongoDB and Cassandra.

Data storage in the cloud is scalable and versatile. You can store files in cloud object storage services like Amazon S3 or Google Cloud Storage, ensuring high availability and durability. These services also support data versioning and access control.

Cloud providers offer identity and access management (IAM) services that allow you to manage users, roles, and permissions for your applications. IAM ensures that only authorized users and services can access your resources, enhancing security.

To secure your applications and data in the cloud, cloud providers offer a wide range of security services and features. These include firewalls, intrusion detection systems, encryption services, and identity verification mechanisms. By taking advantage of these security measures, you can build applications with a strong security posture.

Integration with cloud services often involves APIs and webhooks. APIs allow your application to communicate with cloud services programmatically, enabling actions like sending emails through a cloud-based email service or processing payments via a payment gateway API. Webhooks, on the other hand, allow cloud services to notify your application when specific events occur, such as a new order in an e-commerce system.

When it comes to cloud integration, you have a choice between various communication protocols, including RESTful APIs, GraphQL, and WebSocket. Each protocol has its strengths and is suitable for different scenarios. RESTful APIs, for example, are widely used for their simplicity and compatibility with a wide range of programming languages and frameworks.

As you integrate cloud services into your applications, monitoring and observability become critical. Cloud providers offer tools and services for monitoring application performance, tracking logs, and setting up alerts. These tools help you proactively identify and resolve issues, ensuring a smooth user experience.

Cost management is an important aspect of using cloud services. While the pay-as-you-go model can be cost-effective, it's essential to monitor your resource usage and optimize your spending. Cloud providers offer cost management tools that help you track expenses and identify opportunities for cost savings.

Multi-cloud and hybrid cloud strategies have gained popularity as organizations seek to avoid vendor lock-in and achieve redundancy and fault tolerance. Multi-cloud involves using services from multiple cloud providers, while hybrid cloud combines on-premises infrastructure with cloud resources.

In summary, cloud computing offers a vast landscape of possibilities for building and scaling your applications. By leveraging cloud services and integration options, you can enhance your application's capabilities, improve its scalability, and deliver a robust and feature-rich experience to your users. Whether you're a startup looking for rapid growth or an enterprise seeking to optimize your resources, the cloud is a powerful ally in your quest for success. So, embrace the cloud, explore its services, and let your applications soar to new heights!

Welcome to the exciting world of deploying Python applications to the cloud! In this chapter, we will explore the process of taking your Python code from a local development environment and making it accessible to users worldwide through the power of cloud computing.

Deploying your Python applications to the cloud is an essential step in bringing your software to a broader audience. It allows you to leverage the scalability, reliability, and flexibility of cloud infrastructure, ensuring that your application can handle various workloads and adapt to changing demands.

Before we dive into the specifics of deploying Python applications, it's essential to understand the primary reasons for using the cloud in the deployment process. One of the key advantages of cloud deployment is scalability. Cloud providers offer the ability to scale your application horizontally or vertically, depending on your needs. This means that when your application experiences increased traffic or usage, you can easily allocate more resources to handle the load.

Another significant benefit of cloud deployment is high availability. Cloud providers typically operate multiple data centers in various geographic regions, ensuring redundancy and fault tolerance. This minimizes downtime and enhances the reliability of your application. Users can access your application 24/7, regardless of their location.

Additionally, cloud providers offer managed services that simplify deployment tasks. These services include load balancers, auto-scaling groups, and managed databases. By using these services, you can offload the operational burden of managing infrastructure, allowing you to focus on developing your application's features and functionality.

Now, let's discuss the deployment process itself. When deploying a Python application to the cloud, you have several options. One common approach is to use Platform as a Service (PaaS) offerings provided by cloud providers. PaaS platforms, such as Heroku, Google App Engine, and AWS Elastic Beanstalk, streamline the deployment process by abstracting infrastructure management. You can deploy your Python code without worrying about server provisioning, scaling, or maintenance.

To deploy your Python application to a PaaS platform, you typically need to define a configuration file that specifies your application's dependencies, runtime environment, and any additional settings. Once you've created this configuration, you can use a command-line interface or a web-based dashboard to initiate the deployment process. The PaaS platform takes care of provisioning the necessary infrastructure, deploying your code, and managing runtime environments.

Another deployment option is to use Infrastructure as a Service (IaaS) offerings, such as virtual machines (VMs) provided by cloud providers. With IaaS, you have more control over the underlying infrastructure, allowing you to customize server configurations to meet your specific needs. However, this also means you are responsible for server management tasks, including security updates, patching, and scaling.

When deploying a Python application using IaaS, you typically start by creating a virtual machine instance with the desired operating system. You then configure the virtual machine by installing Python, setting up web servers (e.g., Nginx or Apache), and deploying your application code. Tools like Ansible, Chef, or Puppet can help automate server provisioning and configuration.

Containerization has also gained popularity in the deployment of Python applications. Containers, such as those created with Docker, package your application and its dependencies into a single unit. This unit can be easily deployed and executed in various environments, making it a portable and consistent deployment option.

With containerization, you can build a Docker image that includes your Python code, dependencies, and runtime environment. Once the image is created, you can distribute it to container registries like Docker Hub or Amazon ECR. Cloud providers offer container orchestration services like Kubernetes, which simplify the management and scaling of containerized applications.

Serverless computing is yet another deployment option. In a serverless architecture, you don't need to manage servers or containers. Instead, you focus solely on writing the code that defines the functionality of your application. Cloud providers, such as AWS Lambda, Google Cloud Functions, and Azure Functions, execute your code in response to events or triggers.

Serverless deployment involves defining functions and specifying the events that trigger them. Your Python code is packaged as a function, and the cloud provider handles the underlying infrastructure, including scaling and resource allocation. Serverless deployment is ideal for applications with sporadic workloads and event-driven architectures.

Regardless of the deployment method you choose, it's crucial to consider best practices for securing your Python application in the cloud. This includes implementing access control, encryption, and monitoring solutions to protect your data and infrastructure. Cloud providers offer security services and features that can help you safeguard your application from threats.

To monitor the performance and availability of your deployed application, you can leverage cloud-based monitoring and logging tools. These tools allow you to track application metrics, view logs, and set up alerts for anomalies. By proactively monitoring your application, you can identify and address issues before they impact your users.

In summary, deploying Python applications to the cloud is a vital step in making your software accessible to a global audience. Whether you choose a PaaS platform, IaaS virtual machines, containers, or serverless computing, the cloud offers scalability, reliability, and operational simplicity. By following best practices for security and monitoring, you can ensure that your Python application runs smoothly in the cloud, providing an excellent experience for your users. So, let's embark on this deployment journey and unleash the full potential of your Python creations!

Chapter 6: Building Scalable Web Services with Flask

Welcome to the exciting world of web development with Flask! In this chapter, we'll embark on a journey to explore Flask, a lightweight and flexible web framework for Python. Flask is an excellent choice for building web applications, whether you're creating a simple blog, an e-commerce site, or a powerful web service.

Before we dive into the details of Flask, let's take a moment to understand what a web framework is and why it's essential. In the world of web development, a web framework is like a toolbox that provides developers with pre-built tools, libraries, and conventions to streamline the process of creating web applications. These frameworks handle the low-level details, allowing developers to focus on building the application's unique features and functionality.

Flask, in particular, is known for its simplicity and minimalism. It follows the "micro" framework philosophy, which means it provides only the essentials for web development without imposing rigid structures or unnecessary complexity. This approach makes Flask an excellent choice for developers who prefer flexibility and want to design their applications from the ground up.

One of the first steps in working with Flask is setting up your development environment. To do this, you'll need to install Python if you haven't already. Flask is compatible with both Python 2 and Python 3, but it's recommended to use Python 3 for new projects as Python 2 has reached its end of life.

Once you have Python installed, you can use the Python package manager, pip, to install Flask. Simply open your terminal or command prompt and run the following command:

```
pip install Flask
```

This command will download and install Flask and its dependencies on your system, making it ready for you to start building web applications.

With Flask installed, you can create your first Flask application. But before we jump into coding, let's briefly discuss the structure of a Flask application. In Flask, applications are typically organized into "routes" or "views." Each route corresponds to a URL and defines what should happen when a user visits that URL.

A Flask application typically consists of a Python script, which defines the routes and their associated functions, and a set of HTML templates that determine how the web pages look. Flask allows you to render dynamic content by embedding Python code within the HTML templates.

Here's a basic example of a Flask application that displays a "Hello, World!" message when you visit the root URL ("/"):

pythonCopy code

```
from flask import Flask # Create a Flask application app = Flask(__name__) # Define a route for the root URL @app.route('/') def hello_world(): return 'Hello, World!' # Run the application if __name__ == '__main__': app.run()
```

In this example, we import the Flask class and create an instance of it with **app = Flask(__name__)**. We then define a route using the **@app.route('/')** decorator, which tells Flask to associate the **hello_world** function with the root URL ("/"). When a user visits this URL, the **hello_world** function is executed, and it returns the "Hello, World!" message.

To run the application, we use the **if __name__ == '__main__':** block, which ensures that the application only runs when the script is executed directly, not when it's imported as a module.

Now, let's run the Flask application by executing the script in your terminal or command prompt. You'll see output indicating that the application is running, and you can access it by opening a web browser and navigating to **http://localhost:5000**. You should see the "Hello, World!" message displayed in your browser.

Congratulations! You've just created and run your first Flask application. It's a simple example, but it demonstrates the fundamental concepts of routing and handling requests in Flask.

As you continue your journey with Flask, you'll explore more advanced features, such as rendering templates, handling form submissions, and working with databases. Flask's extensive ecosystem of extensions and plugins will also come in handy as you build more complex web applications.

In the next chapters, we'll delve deeper into Flask's capabilities and cover topics like creating dynamic web pages, handling user input, and building full-fledged web applications. So, get ready to explore the world of Flask, and let's start building amazing web experiences together!

Welcome to the exciting world of web development with Flask! In this chapter, we'll delve into more advanced topics, exploring how to design RESTful APIs and scale your Flask applications to handle a growing number of users and requests.

Understanding RESTful APIs: Before we dive into designing RESTful APIs with Flask, let's briefly understand what a RESTful API is and why it's essential. REST, which stands for Representational State Transfer, is an architectural style for designing networked applications. RESTful APIs, as the name suggests, adhere to these principles.

The REST principles: RESTful APIs are built around a set of key principles:

Resource-Based: In REST, everything is a resource, which can be an object, data, or service. Resources are identified by unique URIs (Uniform Resource Identifiers), and they can be manipulated using standard HTTP methods like GET, POST, PUT, DELETE, etc.

Stateless: Each request to a RESTful API from a client must contain all the information needed to understand and process the request. The server should not rely on any previous requests or sessions to handle the current request.

Representation: Resources can have multiple representations, such as JSON, XML, HTML, or others. Clients can request a specific representation using the **Accept** header in their requests.

Stateless Communication: Communication between client and server is stateless, meaning each request/response cycle is independent. The server does not store information about the client's state between requests.

Designing RESTful APIs with Flask: Flask is an excellent choice for designing RESTful APIs due to its simplicity and flexibility. To create a RESTful API in Flask, you'll need to define routes and map them to functions, similar to traditional web applications. However, instead of rendering HTML templates, your functions will typically return JSON responses. Here's a basic example of a Flask-based RESTful API that exposes a list of books:

pythonCopy code

```
from flask import Flask, jsonify app = Flask(__name__) #
Sample data books = [ {"id": 1, "title": "The Great Gatsby",
"author": "F. Scott Fitzgerald"}, {"id": 2, "title": "To Kill a
Mockingbird", "author": "Harper Lee"}, ]
@app.route('/api/books', methods=['GET']) def
get_books(): return jsonify({"books": books}) if __name__
== '__main__': app.run()
```

In this example, we define a route **/api/books** that responds to HTTP GET requests with a list of books in JSON format. This demonstrates the resource-based nature of RESTful APIs, where books are represented as resources.

Scaling Flask Applications: As your Flask application grows and gains more users, you may need to consider scaling it to handle increased traffic and maintain good performance. Here are some strategies for scaling Flask applications:

Load Balancing: Use load balancers to distribute incoming requests across multiple server instances. This can help distribute the traffic evenly and prevent overloading a single server.

Caching: Implement caching mechanisms to store frequently accessed data and reduce the load on your database or other backend services. Tools like Redis or Memcached are commonly used for caching.

Database Optimization: Optimize database queries and indexes to ensure efficient data retrieval. Use database connection pooling to handle database connections efficiently.

Asynchronous Processing: For tasks that don't need immediate responses, consider using asynchronous processing with libraries like Celery. This can offload time-consuming tasks from the main application thread.

Horizontal Scaling: As your application grows, consider adding more server instances or containers to handle increased load. Tools like Docker and Kubernetes can help manage containerized Flask applications.

Content Delivery Networks (CDNs): Use CDNs to serve static assets like images, CSS, and JavaScript files. CDNs distribute these assets to edge servers closer to users, reducing latency.

Monitoring and Profiling: Implement monitoring and profiling tools to track the performance of your application.

Tools like Prometheus, Grafana, and New Relic can help identify bottlenecks and optimize code.

Auto-Scaling: Set up auto-scaling policies to automatically adjust the number of server instances based on traffic patterns. This ensures that your application can handle spikes in traffic without manual intervention.

Authentication and Security: When designing RESTful APIs, security is paramount. You'll need to implement authentication and authorization mechanisms to protect sensitive data and ensure that only authorized users can access certain resources. Flask provides extensions like Flask-RESTful and Flask-JWT for handling authentication and authorization in API endpoints.

API Documentation: Properly documenting your RESTful API is essential for developers who will be consuming it. Tools like Swagger or Flask-RESTPlus can generate interactive API documentation from your Flask code, making it easier for others to understand and use your API.

Conclusion: Designing RESTful APIs and scaling Flask applications are essential skills for modern web developers. RESTful APIs follow a set of principles that promote simplicity, scalability, and maintainability. Flask's flexibility and ease of use make it a fantastic choice for building RESTful APIs. As your applications grow, scaling strategies like load balancing, caching, and horizontal scaling become crucial to maintaining performance and reliability. Additionally, don't forget to prioritize security, authentication, and documentation to ensure that your API is both functional and secure. So, keep exploring the world of Flask and RESTful APIs, and enjoy building powerful and scalable web applications!

Chapter 7: DevOps and Automation in Python

Welcome to the exciting world of DevOps! In this chapter, we'll explore the fundamental concepts and practices that make up the DevOps philosophy. DevOps is more than just a buzzword; it's a set of principles and practices that aim to bridge the gap between development and operations teams to create a more efficient and collaborative software development process.

What is DevOps: DevOps is a term that combines "Development" and "Operations." It represents a cultural and technical movement that emphasizes collaboration and communication between development and IT operations teams. The goal is to shorten the software development lifecycle and improve the quality and reliability of software deployments.

The Need for DevOps: In traditional software development, development and operations teams often work in silos, leading to challenges such as slow release cycles, manual and error-prone deployments, and a lack of visibility into the production environment. DevOps aims to address these issues by breaking down the barriers between these two groups.

Key Principles of DevOps: DevOps is guided by several key principles:

Collaboration: DevOps encourages collaboration and communication between development, operations, and other relevant teams. This helps in sharing knowledge and responsibilities.

Automation: Automation is at the heart of DevOps. It involves automating repetitive tasks like code integration, testing, deployment, and infrastructure provisioning.

Continuous Integration (CI): CI is the practice of frequently integrating code changes into a shared repository, followed by automated testing and code review. It ensures that code is continuously tested and integrated, reducing integration issues.

Continuous Deployment (CD): CD takes CI a step further by automatically deploying code changes to production or staging environments after passing automated tests. This leads to quicker and more reliable releases.

Monitoring and Feedback: DevOps teams place a strong emphasis on monitoring applications and infrastructure in production. Continuous monitoring provides feedback that can be used to identify and address issues promptly.

Infrastructure as Code (IaC): IaC involves managing infrastructure using code and automation tools. This enables consistent and repeatable infrastructure provisioning.

Benefits of DevOps: Adopting DevOps practices can bring several benefits to organizations:

Faster Release Cycles: DevOps enables faster development, testing, and deployment, allowing organizations to release new features and updates more frequently.

Improved Collaboration: DevOps fosters collaboration between teams, leading to better communication and knowledge sharing.

Reduced Risk: Automation and continuous testing minimize the risk of human errors and ensure that changes are thoroughly validated.

Scalability: IaC and automation make it easier to scale infrastructure up or down as needed.

Enhanced Quality: DevOps practices lead to better-quality software due to automated testing and continuous monitoring.

DevOps Tools: There are a plethora of tools available to support DevOps practices. Some popular ones include Jenkins, Travis CI, GitLab CI/CD, Docker, Kubernetes, Ansible, and Terraform. These tools automate various aspects of the development and deployment process, making it easier to implement DevOps practices.

Challenges of DevOps: While DevOps offers numerous benefits, it also comes with its challenges. Some common challenges include:

Cultural Resistance: Changing the organizational culture to embrace DevOps can be challenging, as it requires breaking down traditional silos and fostering collaboration.

Complexity: Implementing and managing DevOps tools and practices can be complex, especially in large and legacy environments.

Security Concerns: With frequent deployments, security becomes a crucial consideration. DevOps teams must integrate security practices into their workflows.

Skill Gaps: DevOps requires a broad skill set that combines development, operations, and automation expertise. Finding and training individuals with these skills can be a challenge.

DevOps and the Cloud: Cloud computing and DevOps often go hand in hand. Cloud platforms provide the scalability and flexibility needed to support DevOps practices. Cloud services like AWS, Azure, and Google Cloud offer tools and resources that align with DevOps principles.

Conclusion: DevOps is not just a set of tools; it's a cultural shift that promotes collaboration, automation, and continuous improvement. By adopting DevOps practices, organizations can accelerate their development cycles, improve software quality, and respond more effectively to changing market demands. While DevOps may pose challenges, the benefits it brings in terms of speed, quality, and collaboration are well worth the effort. So, whether you're a developer, operations specialist, or a manager, understanding and embracing DevOps can transform the way you approach software development and delivery. Welcome to the exciting realm of automating deployment and infrastructure management. In this chapter, we will delve into the fascinating world of automating the deployment of software applications and managing the underlying infrastructure. This process is a crucial aspect of modern software development and operations, enabling organizations to streamline operations, improve efficiency, and reduce manual errors. At its core, automating deployment and infrastructure management involves using code and automation tools to provision, configure, and manage the infrastructure where your applications run. This practice is often referred to as Infrastructure as Code (IaC), and it revolutionizes the way IT teams handle infrastructure.

Why is automating deployment and infrastructure management essential, you may ask? Well, in the fast-paced world of software development, organizations must respond quickly to changing market demands. Manual infrastructure management can be slow, error-prone, and unable to scale with the pace of development. That's where automation steps in. One of the key concepts in automating deployment and infrastructure management is Infrastructure as Code. In essence, this means that you treat your infrastructure in the same way you treat your application code. You write code that describes your infrastructure, and automation tools use that code to create and manage the actual infrastructure.

So, what benefits does this approach offer? First and foremost, it brings speed and consistency to infrastructure provisioning. With IaC, you can spin up new servers, networking components, and other resources in minutes, reducing the time it takes to respond to business needs. Moreover, the automation ensures that your infrastructure is consistent across different environments, eliminating the dreaded "It works on my machine" problem.

Let's talk about some of the popular tools that facilitate automating deployment and infrastructure management. One of the standout tools in this domain is Terraform. Terraform is an open-source infrastructure as code tool that allows you to define and provision infrastructure across various cloud providers and on-premises data centers. It uses a simple and declarative configuration language, making it accessible to both developers and operations teams. Another notable tool is AWS CloudFormation, specifically tailored for managing resources on the Amazon Web Services (AWS) cloud. With CloudFormation, you can define your AWS infrastructure using templates, enabling you to version-control your infrastructure code and apply changes consistently.

Now, let's explore the concept of Continuous Delivery (CD). CD is an essential practice in automating deployment. It extends the principles of Continuous Integration (CI) by automating the entire delivery process. CI focuses on integrating code changes into a shared repository and running automated tests, while CD takes it a step further, automating the deployment of code to production or staging environments.

CI/CD pipelines are instrumental in automating the deployment process. These pipelines consist of a series of automated steps that include building, testing, and deploying the application. When a developer commits code changes, the CI/CD pipeline kicks off automatically, ensuring that code is built, tested, and deployed consistently.

Containerization is another key concept in automating deployment. Containers, such as those managed by Docker, package applications and their dependencies together. This containerized approach ensures that applications run consistently across different environments, from development laptops to production servers. Kubernetes, an open-source container orchestration platform, has gained significant popularity for managing containerized applications at scale.

Automation also plays a pivotal role in configuration management. Tools like Ansible, Puppet, and Chef automate the configuration and management of servers and applications. These tools allow you to define the desired state of your infrastructure and ensure that it stays in that state, minimizing configuration drift and human errors.

Infrastructure as Code, CI/CD pipelines, and containerization are powerful tools in your automation arsenal, but they are not without their challenges. Ensuring the security of automated infrastructure, managing secrets and credentials, and handling rollback strategies are all critical considerations.

Automation also demands a cultural shift within an organization. Teams must embrace automation as a way to improve their workflows and reduce manual toil. This shift often requires training and a change in mindset but is essential for reaping the full benefits of automation.

In summary, automating deployment and infrastructure management is a transformative practice that can greatly enhance the efficiency and reliability of software development and operations. It allows organizations to respond quickly to changing demands, reduce errors, and maintain consistency across environments. While it comes with challenges, the tools and practices available today make it more accessible than ever. Whether you're a developer, operations specialist, or a manager, understanding and adopting these automation techniques is key to staying competitive in the ever-evolving tech landscape.

Chapter 8: Advanced Python Scripting Techniques

Let's dive into the fascinating world of scripting for performance and efficiency. In this chapter, we'll explore techniques and best practices to make your scripts run faster, consume fewer resources, and become more efficient overall. While writing functional code is essential, optimizing your scripts can significantly enhance their performance, which is especially important in scenarios where speed and efficiency are critical. Performance matters in the world of scripting, whether you're writing small automation scripts or developing complex applications. Users expect software to respond quickly, and inefficient code can lead to frustration and reduced productivity. So, let's roll up our sleeves and uncover some key strategies to script for performance.

One of the foundational principles of scripting for performance is code optimization. This involves identifying and eliminating bottlenecks or inefficiencies in your code. Profiling tools are invaluable in this process; they help you pinpoint which parts of your code are consuming the most time and resources. Once you've identified the bottlenecks, you can focus your optimization efforts where they matter most. Another important consideration is data structures and algorithms. Choosing the right data structures can have a significant impact on the efficiency of your scripts. For instance, using a hash table for fast lookups or a linked list for efficient insertions and deletions can drastically improve the performance of your code. Similarly, optimizing algorithms for the specific task at hand can lead to substantial speed gains. Caching is a powerful technique that can greatly enhance script performance. By storing frequently used data or computation results in memory, you can avoid redundant work. Caching is particularly beneficial for expensive operations, such as database queries or complex calculations. Popular caching mechanisms include in-memory caches like Redis and Memcached.

Parallelism and concurrency are key concepts when it comes to scripting for performance. Parallelism involves executing multiple tasks simultaneously, while concurrency focuses on managing multiple tasks concurrently, even if they don't execute simultaneously. Python's multiprocessing module allows you to harness the power of multi-core processors for parallel execution, while threading provides concurrency for tasks that involve I/O-bound operations.

Furthermore, optimizing I/O operations can significantly boost script performance. When dealing with file I/O, consider using buffered I/O to reduce the number of read and write operations, which can be slow. Additionally, asynchronous I/O can be a game-changer for applications that need to handle many I/O-bound operations concurrently.

Scripting languages like Python offer numerous built-in libraries and modules that can help streamline performance optimization. For instance, the functools module provides decorators like lru_cache for function result caching, while the timeit module helps you measure the execution time of code snippets, aiding in performance testing and optimization.

Speaking of testing, benchmarking your code is crucial for scripting performance. Benchmarking involves running your script under different conditions to measure its performance and identify areas that need improvement. Tools like pytest-benchmark for Python make it easy to set up and run performance tests.

When optimizing your scripts, remember that premature optimization can be counterproductive. Before diving into extensive optimization efforts, profile your code to identify the actual bottlenecks. Optimizing code that doesn't significantly impact performance can lead to reduced code readability and maintainability.

Scripting for performance also involves resource management. Efficiently managing memory, CPU usage, and other system resources is crucial for ensuring that your scripts run smoothly. Resource leaks, such as memory leaks, can degrade script performance over time and lead to unexpected crashes.

Error handling is another aspect of scripting for performance. Poorly implemented error handling can introduce overhead and impact performance negatively. Make sure to handle errors gracefully without adding unnecessary computational burden.

Scripting for performance isn't just about raw speed; it's also about resource efficiency. Minimizing resource usage, whether it's memory, CPU, or network bandwidth, is essential for building efficient and scalable scripts. Resource-efficient scripts are not only faster but also more reliable and cost-effective, especially in cloud-based environments where resource usage directly affects operational costs.

In summary, scripting for performance and efficiency is a crucial skill for developers and scriptwriters. By optimizing your code, choosing the right data structures and algorithms, leveraging caching and parallelism, and effectively managing resources, you can create scripts that not only execute faster but also consume fewer resources. Remember that while performance is important, it should be balanced with code readability and maintainability. Striking this balance will help you create high-quality scripts that meet both functional and performance requirements. So, go ahead and apply these performance optimization techniques to your scripts, and watch them soar to new heights of efficiency.

Chapter 9: Network Automation and Security

Let's delve into the realm of network configuration and automation, my friend. In today's digital age, networks are the backbone of our connected world. They enable communication, data transfer, and access to the internet. Whether you're managing a small home network or a vast enterprise infrastructure, automation can streamline operations and enhance efficiency.

Network automation involves the use of scripts and tools to perform tasks that would traditionally require manual intervention. This can include tasks like configuring network devices, monitoring network performance, and responding to security threats. Automation brings several advantages to the table, such as reducing human errors, saving time, and ensuring consistency across network configurations.

At the heart of network automation lies scripting with programming languages like Python. Python is a versatile language that's well-suited for network automation due to its simplicity and extensive libraries. Python scripts can interact with network devices, gather information, and make configuration changes, all while adhering to best practices and security protocols.

One of the key benefits of using Python for network automation is its ease of integration with networking equipment. Many networking vendors offer Python libraries and APIs (Application Programming Interfaces) that allow you to programmatically interact with their devices. This enables you to automate tasks such as provisioning new network resources, updating configurations, or gathering performance data.

Network automation also plays a crucial role in achieving infrastructure as code (IaC). IaC treats network configurations as code, allowing you to define and manage your network infrastructure using scripts or configuration files. This approach provides version control, consistency, and reproducibility, making it easier to scale and manage complex networks.

Automation in network monitoring is another vital aspect. By using scripts to monitor network performance and health, you can proactively identify issues, receive alerts, and even trigger automated responses. For example, if a network device experiences high CPU utilization, a script can detect this and take corrective actions, such as redistributing traffic to other devices.

Security is a top concern in network management, and automation can help bolster your network's defenses. Security policies and access control lists (ACLs) can be automatically enforced across devices to prevent unauthorized access or malicious activities. Intrusion detection systems (IDS) and intrusion prevention systems (IPS) can also benefit from automation to respond swiftly to threats.

Network documentation is often a neglected aspect of network management, but it's crucial for maintaining an accurate and up-to-date view of your network. Automation can assist in generating network documentation, ensuring that it remains synchronized with the actual network configuration. This documentation includes device inventories, network diagrams, and configuration backups.

Change management is simplified through automation as well. When changes are required in the network, automation scripts can be used to make controlled and consistent modifications. Additionally, automated testing and validation can be applied to ensure that changes do not introduce vulnerabilities or service disruptions.

Scaling network resources up or down is a common requirement, especially in cloud-based environments. With automation, you can dynamically allocate and de-allocate resources based on demand. This elasticity enhances network efficiency and cost-effectiveness.

But with great power comes great responsibility, my friend. Network automation should be approached with care and consideration. It's crucial to thoroughly test automation scripts in a safe environment before deploying them in a production network. This helps prevent unintended consequences or outages due to script errors.

Furthermore, security is paramount in network automation. Ensure that your automation scripts and tools are secure, and implement best practices for managing authentication credentials and access control. Regularly audit and update your automation systems to stay ahead of potential security threats.

In summary, network configuration and automation are pivotal components of modern network management. Python, with its versatility and ease of use, is an ideal choice for scripting and automating network tasks. Whether you're optimizing network performance, enhancing security, or streamlining change management, automation can save you time and reduce the risk of human errors. Embrace the power of automation, but do so responsibly and with a strong focus on security. With the right approach, you can take your network management to the next level and ensure that your network operates smoothly in an ever-evolving digital landscape.

Python has firmly established itself as a powerhouse in the realm of network security and monitoring, empowering professionals to safeguard their digital ecosystems with efficiency and precision. In an era where the integrity of networks and the confidentiality of data are paramount, Python emerges as a versatile ally.

When it comes to network security, vigilance is the name of the game. Threats are omnipresent, ranging from malware and viruses to sophisticated cyberattacks. Python, with its simplicity and rich ecosystem, offers an array of tools and libraries tailored to fortify your network defenses.

In the world of cybersecurity, automation is a game-changer. Python scripts can tirelessly scan networks, analyze logs, and detect anomalies that might signal a breach. They can trigger alerts, isolate compromised devices, and even initiate countermeasures in real-time. With automation, you can respond swiftly to threats, minimizing damage and downtime.

Vulnerability scanning is a cornerstone of network security. Python is an excellent choice for creating custom vulnerability scanners that can pinpoint weaknesses in your network's infrastructure. By regularly scanning and patching vulnerabilities, you can thwart potential exploits before they can be leveraged by malicious actors.

Network traffic analysis is another critical aspect of security. Python allows you to craft custom intrusion detection systems (IDS) and intrusion prevention systems (IPS) that scrutinize network packets for suspicious patterns. These scripts can operate in real-time, monitoring traffic and identifying potential threats, such as intrusion attempts or unauthorized access.

Python's integration capabilities are worth highlighting. Many security appliances and devices offer APIs or Python libraries that facilitate communication and interaction. This means you can orchestrate security tasks like updating firewall rules, configuring access control lists (ACLs), or managing encryption keys seamlessly.

Additionally, Python's prowess in data analysis and visualization is a boon to security professionals. It enables the creation of dashboards and reports that offer insights into network behavior and potential security incidents. By visualizing data, you gain a clearer understanding of your network's strengths and vulnerabilities.

As network security becomes increasingly complex, the importance of log analysis cannot be overstated. Python's extensive libraries for parsing and analyzing log data empower you to extract valuable information, detect patterns, and uncover suspicious activities buried within logs. This analytical capability is invaluable in identifying and mitigating security threats.

Intrusion detection and response (IDR) is an area where Python shines. It enables you to automate the detection of security incidents and initiate predefined responses. For example, when a script detects a brute-force login attempt on a critical server, it can trigger an automatic IP block to thwart further intrusion attempts.

Network forensics, the art of reconstructing events leading to a security incident, is another domain where Python proves its mettle. Python's libraries can sift through vast amounts of data to reconstruct timelines and identify the entry points and origins of security breaches. This forensic capability is essential for incident response and remediation.

When it comes to network monitoring, Python is equally indispensable. It provides the means to track network performance, uptime, and availability. By crafting monitoring scripts, you can ensure that your network operates at peak efficiency, proactively identifying and addressing bottlenecks or performance degradation.

Network monitoring can extend to bandwidth management as well. Python scripts can analyze bandwidth usage, identify bandwidth hogs, and even implement Quality of Service

(QoS) policies to prioritize critical traffic. This ensures that mission-critical applications and services receive the necessary bandwidth for optimal performance.

Network visualization is a powerful tool for monitoring and troubleshooting. Python libraries like Matplotlib and NetworkX allow you to create visual representations of your network's topology, helping you grasp its complexity and dependencies. Visualization aids in identifying potential single points of failure and optimizing network design.

An often-overlooked aspect of network security and monitoring is compliance. Regulatory requirements demand adherence to specific security standards. Python can assist in automating compliance checks, ensuring that your network configurations and security policies align with industry regulations.

While Python empowers you to create custom security and monitoring solutions, it's crucial to consider best practices. Security scripts themselves must be secure, and access to them should be restricted to authorized personnel. Encryption and secure communication protocols should be employed when exchanging sensitive data.

In summary, Python is a formidable ally in the domains of network security and monitoring. Its versatility, automation capabilities, and rich ecosystem of libraries make it an invaluable tool for fortifying your network defenses and maintaining vigilance against evolving threats. By harnessing the power of Python, you can safeguard your network and data with confidence in an ever-changing digital landscape.

Chapter 10: Advanced Automation Projects and Real-World Applications

Automation, powered by Python's versatility and adaptability, has permeated virtually every sector, reshaping processes, and making them more efficient and scalable. In this chapter, we delve into real-world use cases where advanced automation plays a pivotal role, illustrating how Python becomes a transformative force.

Let's begin our journey in the realm of finance. Financial institutions handle enormous volumes of data daily, from trading information to customer transactions. Python's data manipulation and analysis libraries, such as Pandas and NumPy, shine in this space. Traders and analysts rely on Python to automate the extraction, cleansing, and analysis of data, providing insights that inform investment decisions. Algorithmic trading, a prime example, employs Python to implement complex trading strategies that respond swiftly to market fluctuations. The result? Enhanced efficiency and better-informed investment decisions.

Moving to healthcare, we find Python at the forefront of medical image analysis. Radiologists leverage Python to create algorithms that assist in the diagnosis of conditions like cancer and fractures. By automating the analysis of medical images, Python helps healthcare providers save time and reduce human error, ultimately improving patient outcomes.

The e-commerce landscape is another domain where Python automation thrives. Online retailers, for instance, employ Python to optimize their supply chain management. By automating inventory tracking and order fulfillment, businesses ensure they have the right products in stock and that orders are processed and shipped promptly. This not only enhances customer satisfaction but also boosts operational efficiency and reduces costs.

Python's applications in the entertainment industry are equally fascinating. Streaming platforms, such as Netflix and Spotify, employ Python for recommendation systems. These systems analyze user preferences and behavior to recommend content tailored to individual tastes. Python's ability to process vast datasets and apply machine learning models is instrumental in creating a personalized user experience.

Now, let's journey into the world of manufacturing. Automation has revolutionized production lines, with Python serving as a programming language of choice for control systems. Python's simplicity and readability make it an ideal tool for developing and maintaining automation scripts that manage everything from robotics to quality control. In manufacturing, Python-driven automation ensures consistency, precision, and rapid adaptation to changing production needs.

Moving on to the transportation sector, Python is instrumental in optimizing logistics and route planning. Delivery companies employ Python to create algorithms that determine the most efficient routes for their fleets, considering factors like traffic, weather, and delivery windows. This not only reduces fuel consumption but also enhances customer satisfaction through on-time deliveries.

The energy industry benefits from Python's automation capabilities as well. Power plants and utilities utilize Python to monitor and manage energy consumption in real-time. By automating the analysis of data from smart meters and sensors, Python helps optimize energy distribution, reduce waste, and enhance the stability of the electrical grid.

Python's influence extends to the field of scientific research, where it plays a critical role in data analysis and simulation.

Researchers in various disciplines, from physics to biology, leverage Python to process and visualize data, conduct simulations, and develop models that advance scientific understanding. Python's extensive libraries and the scientific ecosystem, including SciPy, NumPy, and Matplotlib, empower researchers to automate complex scientific tasks efficiently.

In the realm of agriculture, Python-driven automation assists farmers in optimizing crop management. Sensors and drones equipped with Python scripts gather data on soil conditions, crop health, and weather patterns. By automating data analysis, farmers can make informed decisions about irrigation, fertilization, and pest control, ultimately increasing crop yields and reducing resource waste.

Python also has a significant impact on education. Instructors and educational institutions employ Python to automate administrative tasks, such as grading assignments and managing student records. Additionally, Python's accessibility and simplicity make it an excellent choice for teaching programming and computational thinking to students of all ages.

Cybersecurity is another area where Python shines. Security professionals use Python to automate vulnerability scanning, intrusion detection, and incident response. By continuously monitoring networks and automating threat detection, organizations can bolster their defenses against cyberattacks and respond swiftly to security incidents.

In the realm of customer service, Python-powered chatbots are becoming increasingly prevalent. These chatbots use natural language processing (NLP) libraries to understand and respond to customer inquiries. By automating routine customer interactions, organizations can provide efficient and round-the-clock support while freeing up human agents to focus on more complex tasks.

Python's impact on the environment is noteworthy as well. Environmental monitoring stations rely on Python to collect and analyze data on air quality, weather, and ecological trends. This data informs policymakers and researchers about the state of the environment and helps develop strategies for conservation and sustainable practices.

In summary, Python's role in advanced automation is nothing short of transformative. From finance to healthcare, e-commerce to manufacturing, and beyond, Python's versatility and adaptability empower professionals across diverse industries to automate processes, make data-driven decisions, and achieve new levels of efficiency and innovation. As technology continues to evolve, Python remains a dynamic tool in the hands of those who seek to push the boundaries of what automation can achieve.

Implementing automation in various domains is a journey that involves harnessing the power of technology to streamline processes and enhance productivity. It's about finding innovative ways to leverage automation tools and solutions to achieve specific goals and drive positive outcomes. Let's explore how automation is making a significant impact in different domains.

In the world of business, automation is a game-changer. Organizations of all sizes and across industries are embracing automation to optimize their operations. This includes automating routine administrative tasks, such as data entry and report generation, which frees up employees to focus on more strategic and creative aspects of their roles. Moreover, automation is helping companies gain a competitive edge by improving efficiency, reducing costs, and enhancing customer experiences.

Customer relationship management (CRM) is one area where automation shines. Businesses use CRM software to automate customer interactions, track leads, and manage sales pipelines. Automated email marketing campaigns, for instance, allow organizations to reach out to their customers with personalized content at scale, increasing engagement and driving sales.

In the field of marketing, automation plays a pivotal role in lead generation and nurturing. Marketing automation platforms enable businesses to create automated workflows that send targeted messages to prospects based on their behavior and preferences. This not only saves time but also ensures that leads receive timely and relevant information, increasing the likelihood of conversion.

Moving into the healthcare domain, automation is transforming patient care. Electronic health records (EHR) systems automate the storage and retrieval of patient data, reducing the risk of errors and enabling healthcare providers to make more informed decisions. Moreover, automation is being used to schedule appointments, send medication reminders, and even assist with telemedicine services, making healthcare more accessible and efficient.

In the manufacturing sector, automation has led to significant advancements in production processes. Robotics and industrial automation systems are used to perform tasks ranging from assembly to quality control with precision and speed. This not only increases output but also improves product quality and safety for workers.

Supply chain management is another area where automation is driving efficiency. Automated inventory management systems help businesses keep track of stock levels in real-time, allowing for just-in-time inventory practices that reduce carrying costs. Moreover, automation is being used for demand forecasting, optimizing shipping routes, and even monitoring the condition of goods during transit.

The financial sector has long been a pioneer in adopting automation. Automated trading algorithms execute buy and sell orders in milliseconds, responding to market changes far faster than human traders ever could. Additionally, fraud detection systems use automation to analyze transactions and flag suspicious activities, helping banks and financial institutions protect their customers' assets.

Education is not untouched by the automation wave. Educational institutions are using learning management systems (LMS) to automate the delivery of course materials, quizzes, and assessments. These platforms facilitate remote learning, making education accessible to a global audience. Furthermore, chatbots and virtual teaching assistants help answer student queries and provide instant feedback.

When it comes to agriculture, precision farming is revolutionizing the industry. Automated machinery equipped with sensors and GPS technology can plant, harvest, and fertilize crops with incredible precision. Farmers can also monitor soil conditions and weather forecasts in real-time, allowing them to make data-driven decisions that maximize crop yields and minimize resource usage.

Automation has also made its mark in the realm of entertainment and content creation. Streaming platforms use recommendation algorithms to suggest content to users based on their viewing habits. Additionally, automated video editing tools can transform raw footage into polished videos, making content creation more accessible to creators with varying levels of expertise.

The transportation and logistics sector benefits from automation in numerous ways. Automated vehicle tracking systems help companies monitor the location and condition of their fleets, ensuring timely deliveries and reducing fuel consumption. In warehouses, robots work alongside human employees to pick and pack orders efficiently.

In summary, implementing automation in various domains has become a hallmark of progress and innovation. From optimizing business operations to enhancing customer experiences, improving healthcare to revolutionizing manufacturing, automation is a force that continues to shape the way we work and live. As technology evolves, so too will the possibilities for automation, offering new opportunities for efficiency, creativity, and growth across industries.

BOOK 4
PYTHON AUTOMATION MASTERY
EXPERT-LEVEL SOLUTIONS

ROB BOTWRIGHT

Chapter 1: Mastering Design Patterns in Python

Creational design patterns are fundamental in software development, and they play a crucial role in object-oriented programming. These patterns provide solutions to the problem of object creation, and they help ensure that objects are created in a way that is both flexible and efficient.

One of the most commonly used creational design patterns is the Singleton pattern. It ensures that a class has only one instance and provides a global point of access to that instance. This can be particularly useful in situations where you want to restrict the instantiation of a class to just one object, such as managing a shared resource or configuration settings.

Another important creational pattern is the Factory Method pattern. It defines an interface for creating an object, but it allows subclasses to alter the type of objects that will be created. This promotes loose coupling between the creator and the product, making it easier to extend and maintain code.

The Abstract Factory pattern takes the concept of the Factory Method pattern a step further. It provides an interface for creating families of related or dependent objects without specifying their concrete classes. This is especially valuable when you need to ensure that the created objects are compatible and work together seamlessly.

Builder pattern is a creational pattern that is often used to construct a complex object step by step. It separates the construction of a complex object from its representation, allowing the same construction process to create different representations. This can be beneficial when dealing with complex objects with many optional parts or configurations.

Prototype pattern involves creating new objects by copying an existing object, known as the prototype. This can be advantageous when the cost of creating an object is more expensive than copying an existing one. It allows you to create new objects with minimal overhead by cloning an existing instance.

Creational design patterns are not limited to single-object creation. They can also be applied to creating families of related objects. The Factory Method and Abstract Factory patterns are examples of this. They enable the creation of objects with a common theme or purpose, ensuring that they work together seamlessly within a system.

When deciding which creational design pattern to use, it's essential to consider the specific requirements and constraints of your project. Each pattern has its strengths and weaknesses, and the choice often depends on the complexity of the objects you need to create and how they relate to each other.

In practice, creational design patterns are not used in isolation. They are often combined with other design patterns, such as structural and behavioral patterns, to address complex software design challenges comprehensively. This approach allows developers to build robust, maintainable, and flexible software systems.

To illustrate the importance of creational design patterns, let's consider a real-world example. Imagine you are developing a game that features various types of characters. Each character has specific attributes and abilities. Without creational design patterns, you might end up with a cluttered and error-prone codebase.

By applying the Factory Method pattern, you can create a factory for each type of character, such as a

"WarriorFactory" and a "WizardFactory." These factories encapsulate the logic for creating characters of their respective types, ensuring that the construction process is consistent and maintainable.

Moreover, if you decide to add a new type of character, you can easily extend your system by creating a new factory class, without needing to modify existing code. This demonstrates how creational design patterns promote code reusability and extensibility, making them invaluable tools in software development.

In summary, creational design patterns are essential elements of object-oriented software design. They provide flexible and efficient solutions to the problem of object creation, promoting code reusability, maintainability, and extensibility. By understanding and applying these patterns, software developers can build more robust and scalable systems, enhancing the overall quality of their software projects.

Structural and behavioral design patterns are two fundamental categories in the world of software design patterns. They offer solutions to common problems that developers encounter when designing and implementing software. These patterns help make code more organized, maintainable, and adaptable, ultimately leading to more robust and efficient software systems.

Structural Design Patterns

Structural design patterns focus on the composition of classes and objects. They deal with how objects can be assembled to form larger, more complex structures while keeping the system flexible and easy to modify. Let's explore some essential structural design patterns.

Adapter Pattern: This pattern allows the interface of an existing class to be used as another interface, making it compatible with other classes. It's like a translator between two different systems that wouldn't otherwise work together.

Bridge Pattern: The bridge pattern separates an object's abstraction from its implementation, allowing them to vary independently. This is particularly useful when you want to avoid a permanent binding between an abstraction and its implementation.

Composite Pattern: The composite pattern lets you compose objects into tree structures to represent part-whole hierarchies. It treats individual objects and compositions of objects uniformly, making it easier to work with complex structures.

Decorator Pattern: Decorators are used to add responsibilities to objects dynamically. They provide a flexible alternative to subclassing for extending functionality. For example, you can add new behaviors to an object without altering its class.

Facade Pattern: The facade pattern provides a simplified interface to a set of interfaces in a subsystem, making it easier to use. It acts as a gateway to a more complex system, shielding clients from its complexity.

Flyweight Pattern: This pattern minimizes memory usage or computational expenses by sharing as much as possible with related objects. It's particularly useful when dealing with a large number of similar objects.

Proxy Pattern: A proxy is a placeholder for another object to control access to it. It can be used for various purposes, such as lazy loading, access control, or monitoring.

Behavioral Design Patterns

Behavioral design patterns, on the other hand, deal with the interaction and responsibility of objects. They focus on how objects collaborate and communicate with each other. Here are some key behavioral design patterns.

Chain of Responsibility Pattern: This pattern passes a request along a chain of handlers, with each handler deciding whether to process the request or pass it to the next handler. It promotes loose coupling between senders and receivers.

Command Pattern: The command pattern encapsulates a request as an object, allowing you to parameterize clients with queues, requests, and operations. It also provides support for undoable operations.

Interpreter Pattern: Interpreter is used to define a grammar for interpreting a language and provides an interpreter to interpret sentences in that language. It's often used in language processing.

Iterator Pattern: The iterator pattern provides a way to access the elements of an aggregate object sequentially without exposing its underlying representation. It simplifies the traversal of collections.

Mediator Pattern: Mediator defines an object that centralizes communication between other objects. It promotes a more structured way of communication and reduces direct connections between components.

Memento Pattern: The memento pattern captures an object's internal state and allows it to be restored to that state. It's commonly used for undo mechanisms.

Observer Pattern: The observer pattern defines a one-to-many dependency between objects, so when one object changes state, all its dependents are notified and updated automatically. It's widely used in event handling systems.

State Pattern: The state pattern allows an object to alter its behavior when its internal state changes. It encapsulates states into separate classes and delegates the behavior to the current state object.

Strategy Pattern: The strategy pattern defines a family of algorithms, encapsulates each one, and makes them interchangeable. It allows you to select an algorithm's behavior at runtime.

Template Method Pattern: The template method defines the skeleton of an algorithm in a method, deferring some steps to subclasses. It provides a way to define the structure of an algorithm while allowing its specific steps to be implemented by subclasses.

Visitor Pattern: The visitor pattern represents an operation to be performed on the elements of an object structure. It lets you define a new operation without changing the classes of the elements on which it operates.

Understanding these structural and behavioral design patterns is crucial for designing and developing maintainable and scalable software. They provide solutions to common design challenges and guide developers toward creating code that's not only functional but also easy to maintain, extend, and adapt to changing requirements.

When deciding which design pattern to apply, it's essential to consider the specific problem you're trying to solve and the context in which it arises. Different patterns address different aspects of software design, and choosing the right one can significantly impact the quality and maintainability of your code.

In practice, you'll often find that combining multiple design patterns is necessary to build complex software systems effectively. For example, you might use the Observer pattern to handle events and notifications in your application while employing the Factory pattern to create objects with specific behaviors. As you gain experience in software development, you'll develop a deeper understanding of when and how to apply these patterns effectively. They become valuable tools in your toolkit, allowing you to tackle a wide range of design challenges and build robust and scalable software systems.

Chapter 2: Performance Optimization and Profiling

Identifying performance bottlenecks in your software is a crucial aspect of software development. It's like being a detective, searching for clues to solve a mystery. When your software runs slowly or doesn't meet its performance goals, it can be frustrating for both developers and users. In this chapter, we'll explore strategies and techniques to help you uncover and address performance bottlenecks effectively.

First, let's understand what a performance bottleneck is. Imagine a highway during rush hour. Traffic is slow, and cars are moving at a snail's pace. The bottleneck, in this case, is a constriction in the road that limits the flow of traffic, causing delays. In software, a performance bottleneck is a part of your code or system that hinders overall performance. It's like a traffic jam in your application.

Performance bottlenecks can manifest in various ways. Your software might be slow to respond to user interactions, take too long to process data, or struggle to handle a high volume of concurrent users. These issues can lead to poor user experiences and impact your application's usability and scalability.

So, how do you become a performance detective and identify these bottlenecks? The first step is to establish performance goals for your application. What response times do you want to achieve? How many concurrent users should your system support? Having clear goals will guide your investigation.

Once you have goals in place, it's time to gather data. Profiling tools are your investigative tools of choice. These tools help you collect data about your application's behavior, such as which functions or methods take the most time to execute, how much memory is being used, and where your code spends the most time.

Profiling tools come in various forms, from built-in Python modules like cProfile and memory_profiler to third-party tools like Pyflame, which can help you identify CPU bottlenecks. For web applications, tools like Django Debug Toolbar or Flask Debug Toolbar provide valuable insights into request-response cycles.

With data in hand, it's time to analyze the results. Look for patterns and anomalies in your profiling data. Are there specific functions or methods that consistently consume a lot of CPU time or memory? These might be the culprits behind your performance issues.

But don't jump to conclusions too quickly. Sometimes, the code consuming the most resources isn't the root cause of the problem. It could be a symptom of a deeper issue. Take the time to understand the context in which your code is executing. Are there unexpected loops or recursive calls? Are you making redundant database queries or network requests?

Profiling data might reveal that database queries or I/O operations are slowing down your application. In such cases, consider optimizing your queries, using caching mechanisms, or employing asynchronous programming to reduce waiting times.

Concurrency issues can also be a common source of bottlenecks, especially in multi-threaded or multi-process applications. Deadlocks, contention for resources, or inefficient thread synchronization can lead to performance problems. Tools like Python's **threading** and **multiprocessing** modules offer ways to address these issues.

Once you've identified the bottlenecks, it's time to implement optimizations. This might involve rewriting or refactoring specific parts of your code, using more efficient algorithms, or leveraging caching mechanisms. Keep in mind that premature optimization can be counterproductive, so focus on addressing the most critical bottlenecks first.

Performance testing is another essential aspect of identifying bottlenecks. Load testing tools like Apache JMeter or locust.io can simulate high levels of traffic to your application, helping you understand how it performs under stress. Monitoring tools like Prometheus and Grafana provide real-time insights into your application's performance in production environments.

A proactive approach to performance optimization involves continuous monitoring and testing. Set up alerts and automated tests to detect performance regressions early in the development process. This way, you can catch bottlenecks before they impact your users.

It's also worth mentioning that performance optimization is an iterative process. As your application evolves and scales, new bottlenecks may emerge. Regularly revisit your performance goals, gather fresh profiling data, and make refinements to your code and infrastructure as needed.

Remember that performance optimization is not a one-size-fits-all endeavor. The techniques and tools you use will depend on your application's specific requirements and technologies. What works for a web application might not apply to a data processing pipeline or a real-time game engine. In summary, becoming a performance detective in software development involves setting clear performance goals, using profiling and monitoring tools to gather data, analyzing the results, and implementing optimizations based on your findings. It's an ongoing process that requires attention to detail and a commitment to delivering a fast and responsive user experience. By identifying and addressing performance bottlenecks, you can ensure that your software runs smoothly and efficiently, delighting users and stakeholders alike. Optimizing Python code is like fine-tuning a musical instrument; it requires a keen ear and a deep understanding of the performance bottlenecks that may arise in your software. While Python is a versatile and expressive language, it can sometimes be slower compared to compiled languages like C or Java. However, with the right profiling tools and optimization techniques, you can achieve impressive speed improvements.

Let's start by discussing the importance of profiling. Profiling is like putting your code under a microscope. It enables you to examine its behavior in fine detail, identifying which parts are consuming the most time and resources. Python provides built-in profiling modules like **cProfile** and **profile**, which are invaluable for understanding your code's performance.

When you profile your code, you'll often encounter two primary types of bottlenecks: CPU-bound and memory-bound. CPU-bound bottlenecks occur when your code spends too much time processing computations, while memory-bound bottlenecks arise when your code consumes excessive memory.

To address CPU-bound bottlenecks, consider the following strategies. First, focus on algorithmic improvements. Sometimes, a more efficient algorithm can dramatically reduce execution time. For instance, replacing a nested loop with a more optimized algorithm can be a game-changer.

Another approach is to use built-in Python functions and libraries that are implemented in C. Functions like **numpy** for numerical operations or **itertools** for iterators are much faster than equivalent pure Python code. Leveraging these can lead to significant speed gains.

Parallelism and concurrency are also powerful tools for CPU-bound tasks. Python's **multiprocessing** and **concurrent.futures** modules can help you distribute work across multiple CPU cores or execute tasks asynchronously, reducing execution time.

For memory-bound bottlenecks, memory profiling tools like **memory_profiler** are essential. They provide insights into memory consumption and help identify memory leaks or inefficient data structures.

Optimizing memory usage involves techniques like lazy loading, which loads data into memory only when needed, or using data structures that have lower memory overhead. The **collections** module in Python offers memory-efficient alternatives to standard data structures.

Caching is another strategy for improving memory usage. By storing and reusing computed results, you can reduce redundant calculations and save memory. Libraries like **cachetools** can simplify caching implementation.

When working with large datasets, consider memory-mapped files, which allow you to map a file directly into memory, avoiding the need to load it entirely into RAM. The **mmap** module in Python facilitates this technique.

Beyond optimizing code, profiling can reveal performance bottlenecks related to I/O operations. Disk and network operations can slow down your application significantly. Caching and asynchronous programming are valuable strategies for handling I/O-bound tasks.

Sometimes, the most effective optimization is to avoid doing unnecessary work. For example, if you're fetching data from a web API, check if the data has changed since the last request before fetching it again. This can save precious network bandwidth and processing time.

In web applications, optimizing database queries is essential. Use database indexes, limit the number of rows retrieved, and employ pagination to reduce database load. Tools like Django Debug Toolbar or SQLAlchemy's query profiler can help you analyze query performance.

One advanced optimization technique is Just-In-Time (JIT) compilation. Libraries like **Numba** allow you to compile Python code to machine code at runtime, achieving near-native execution speed. However, JIT compilation is most beneficial for computationally intensive tasks.

Lastly, don't forget about the importance of testing after optimizations. Automated tests can catch regressions and ensure that your optimizations haven't introduced new bugs. Continuous integration and continuous deployment (CI/CD) pipelines can help streamline this process.

In the world of Python optimization, the key is balance. It's important to strike a balance between readable, maintainable code and code optimized for performance. Premature optimization can make code less readable and harder to maintain, so focus on optimizing critical sections of your codebase.

Remember that profiling and optimization are not one-time tasks. As your code evolves and your application scales, new bottlenecks may emerge. Regularly profiling and optimizing your code can help maintain its performance over time.

In summary, profiling and optimizing Python code involve a combination of techniques, including algorithmic improvements, leveraging built-in libraries, parallelism, and smart memory management. Profiling tools and memory profiling help you pinpoint bottlenecks, and optimization strategies can be tailored to address CPU-bound, memory-bound, or I/O-bound challenges. Striking a balance between code readability and performance is key, and regular testing ensures that optimizations don't introduce new issues. With these tools and techniques, you can achieve remarkable speed improvements in your Python applications.

Chapter 3: Advanced Python Concurrency and Parallelism

Concurrency and parallelism are two terms often used in the context of multithreading and multiprocessing, and they play a crucial role in improving the performance and responsiveness of software systems. While these terms are related, they refer to different concepts and are used in distinct scenarios.

Concurrency is like juggling multiple tasks in the air, where you switch between tasks quickly to give the illusion of simultaneous execution. In software, concurrency refers to the ability of a program to manage multiple tasks concurrently, even if they don't run simultaneously. Concurrency is typically achieved through techniques like multitasking, multithreading, or asynchronous programming. Imagine a web server handling multiple client requests. Concurrency allows the server to accept and process incoming requests concurrently, giving the impression that they are being served simultaneously. In reality, the server is rapidly switching between different requests.

Python's Global Interpreter Lock (GIL) is a unique feature that can impact concurrency. The GIL allows only one thread to execute Python bytecode at a time in a single process, making it challenging to fully utilize multicore processors for CPU-bound tasks. However, concurrency is still valuable for I/O-bound tasks in Python because it enables a single thread to handle multiple I/O operations efficiently.

On the other hand, parallelism is akin to having multiple workers performing tasks simultaneously. In the world of software, parallelism refers to the actual simultaneous execution of multiple tasks, often leveraging multiple CPU cores or processors. It's about splitting a large task into smaller subtasks that can be executed concurrently.

Consider an image processing application that applies filters to a large set of images. Parallelism allows the application to distribute the work across multiple CPU cores, processing several images at the same time. This significantly speeds up the overall processing time.

In Python, the **multiprocessing** module provides tools for achieving parallelism by creating multiple processes, each with its own Python interpreter and memory space. This allows Python to utilize multiple CPU cores effectively, overcoming the limitations imposed by the GIL.

To summarize, concurrency focuses on managing and scheduling tasks to make the most efficient use of available resources, even when tasks don't truly run simultaneously. It's valuable for I/O-bound tasks and can improve the responsiveness of software. Parallelism, on the other hand, involves executing tasks simultaneously, often using multiple CPU cores, and is particularly beneficial for CPU-bound tasks. The choice between concurrency and parallelism depends on your specific use case. If your application performs a mix of I/O-bound and CPU-bound tasks, combining concurrency and parallelism can be a winning strategy. For I/O-bound tasks, you can leverage concurrency to handle multiple tasks simultaneously without waiting for I/O operations to complete. For CPU-bound tasks, parallelism allows you to fully utilize the processing power of your hardware.

It's important to note that both concurrency and parallelism come with challenges. Coordinating concurrent tasks to avoid race conditions and deadlocks can be complex, and it requires careful synchronization. In the case of parallelism, you need to ensure that tasks can be divided into independent units of work and that the overhead of managing multiple processes doesn't outweigh the performance gains.

In summary, concurrency and parallelism are powerful techniques for improving the performance and responsiveness of software. Concurrency enables efficient task scheduling, making the most of available resources, while parallelism achieves true simultaneous execution, particularly beneficial for CPU-bound tasks. The choice between them depends on your application's specific requirements, and combining both approaches can be a winning strategy for many real-world scenarios.

In the ever-evolving world of software development, the need for efficient and responsive applications is paramount. Whether you're building a web server, a data processing pipeline, or a desktop application, concurrency is a crucial concept that can help you make the most of your computing resources. Python, known for its simplicity and versatility, offers several libraries and techniques to implement concurrency effectively.

Threading:

Python's **threading** module provides a high-level interface to create and manage threads within a single process. Threads are lightweight, and they share the same memory space, making them suitable for tasks that involve I/O operations or other tasks where waiting for external resources is common. Python's Global Interpreter Lock (GIL) can limit the benefits of multithreading for CPU-bound tasks, but it doesn't diminish the advantages of concurrency for I/O-bound tasks.

Imagine you're building a web server that needs to handle multiple client requests simultaneously. Threading allows you to create a new thread for each incoming request, ensuring that the server can serve multiple clients concurrently. This responsiveness is crucial for delivering a smooth user experience.

Multiprocessing:
While threading is suitable for I/O-bound tasks, Python's **multiprocessing** module shines when dealing with CPU-bound tasks. Multiprocessing overcomes the GIL limitations by creating separate processes, each with its own Python interpreter and memory space. This means that multiple CPU cores can be fully utilized to execute tasks simultaneously.

Consider a scenario where you're running a complex simulation or data analysis that involves intensive computations. Multiprocessing allows you to split the workload across multiple processes, taking advantage of the full processing power of your hardware.

Asyncio:
Python's **asyncio** library introduces asynchronous programming, which is a powerful way to handle concurrency for I/O-bound tasks, particularly in networked applications. Instead of using threads or processes, asyncio employs coroutines and an event loop to efficiently manage asynchronous tasks. It allows you to write non-blocking, highly concurrent code that can handle many simultaneous I/O operations without the overhead of thread or process creation.

Imagine you're developing a web crawler or a chat application where numerous network requests need to be made concurrently. Asyncio enables you to write efficient, event-driven code that can handle these tasks concurrently, ensuring responsiveness and scalability.

Concurrency Patterns:
In addition to the libraries mentioned, Python also supports various concurrency patterns, such as the producer-consumer pattern, the thread pool pattern, and the worker pool pattern. These patterns provide higher-level

abstractions for managing concurrent tasks, making it easier to implement common concurrency scenarios.

For example, in a producer-consumer pattern, you can have multiple threads or processes producing data and multiple threads or processes consuming and processing that data. This pattern is handy for scenarios like data streaming or data transformation pipelines.

Concurrency Challenges:

While concurrency offers significant benefits, it also comes with its share of challenges. Managing shared resources, avoiding race conditions, and ensuring thread safety are crucial concerns when dealing with concurrent code. Python provides synchronization mechanisms like locks, semaphores, and conditions to address these challenges.

Moreover, debugging concurrent programs can be complex because issues may not manifest consistently and can be challenging to reproduce. Tools like thread and process debugging, along with specialized libraries like **asyncio**'s **asyncio.gather()**, are invaluable for identifying and resolving concurrency-related bugs.

Choosing the Right Approach:

Deciding which concurrency technique to use depends on your specific use case. If your application involves I/O-bound operations, consider using threading or asyncio. For CPU-bound tasks, multiprocessing is a better fit. It's also common to combine these techniques in a single application to leverage their strengths where they matter most.

Chapter 4: Cybersecurity and Ethical Hacking with Python

In today's increasingly interconnected world, where digital information is the lifeblood of businesses and individuals alike, cybersecurity is of paramount importance. With cyberattacks becoming more sophisticated and prevalent, ethical hacking has emerged as a crucial defense mechanism. Ethical hackers, also known as "white hat" hackers, play a pivotal role in identifying vulnerabilities, fortifying systems, and safeguarding sensitive data. In this chapter, we will delve into the fundamentals of ethical hacking, exploring its significance, methodologies, and ethical considerations.

The Significance of Ethical Hacking:

Cybersecurity breaches can have catastrophic consequences, ranging from financial losses and reputational damage to the compromise of sensitive information. Ethical hacking serves as a proactive approach to identifying and rectifying vulnerabilities before malicious hackers exploit them. By mimicking the techniques and tactics employed by malicious hackers, ethical hackers can help organizations fortify their digital defenses and ensure the integrity and confidentiality of their data.

Methodologies in Ethical Hacking:

Ethical hacking is a structured and systematic process that involves several key phases. The following are the primary stages in the ethical hacking methodology:

Reconnaissance: Ethical hackers begin by gathering information about the target, including its network architecture, system configurations, and potential vulnerabilities. This phase relies on open-source intelligence, publicly available information, and passive scanning techniques.

Scanning: In this phase, ethical hackers use active scanning tools and techniques to identify vulnerabilities and weaknesses in the target systems. They may employ network scanners, port scanners, and vulnerability scanners to assess the security posture of the target.

Enumeration: Enumeration involves extracting additional information about the target, such as user accounts, system resources, and network services. Ethical hackers aim to gain a deeper understanding of the target's environment to identify potential entry points.

Vulnerability Analysis: During this phase, ethical hackers meticulously analyze the information gathered in the previous steps to identify specific vulnerabilities and weaknesses in the target's infrastructure. Vulnerability databases and security advisories play a crucial role in this process.

Exploitation: Once vulnerabilities are identified, ethical hackers attempt to exploit them to gain unauthorized access to systems or applications. This phase helps organizations understand the potential impact of a successful attack and assess their overall security.

Post-Exploitation: After gaining access, ethical hackers aim to maintain their presence and assess the extent of potential damage. They may also pivot to other systems or escalate privileges to understand the full scope of the vulnerabilities.

Reporting: The final phase involves compiling a comprehensive report that outlines the vulnerabilities discovered, their potential impact, and recommendations for remediation. This report is then shared with the organization's stakeholders to facilitate mitigation efforts.

Ethical Considerations:

Ethical hacking is guided by a strict code of ethics designed to ensure that the process remains legal, responsible, and aligned with the best interests of the organization being tested. Ethical hackers are committed to:

Legal Compliance: Ethical hackers must operate within the boundaries of the law. They should obtain proper authorization and consent before conducting any security assessments.

Confidentiality: Ethical hackers are entrusted with sensitive information about an organization's security posture. They are obligated to protect this information and not disclose it without explicit consent.

Integrity: Ethical hackers must conduct their assessments with the utmost integrity, avoiding any actions that may cause harm or disrupt the organization's operations.

Responsible Disclosure: When vulnerabilities are discovered, ethical hackers follow a responsible disclosure process. This involves notifying the affected organization promptly and providing them with sufficient time to address the issues before making the findings public.

Continuous Learning and Adaptation:

The field of ethical hacking is dynamic and ever-evolving. New attack vectors and vulnerabilities emerge regularly, requiring ethical hackers to stay current with the latest threats and mitigation strategies. Continuous learning is essential, and ethical hackers often engage in training, certifications, and participation in the broader cybersecurity community to hone their skills and knowledge.

Tools and Techniques:

Ethical hackers leverage a wide array of tools and techniques to assess the security of systems and networks. These tools include network scanners like Nmap, vulnerability scanners like Nessus, password cracking tools like John the Ripper, and exploitation frameworks like Metasploit. Additionally, ethical hackers may employ social engineering tactics, phishing simulations, and physical security assessments to comprehensively evaluate an organization's security posture.

Ethical Hacking in Practice:

Ethical hacking is employed across various industries and sectors. Organizations of all sizes, from startups to multinational corporations, recognize the importance of proactive security testing. Ethical hackers may work in-house as part of an organization's cybersecurity team, as independent consultants, or as members of specialized security firms.

In summary, ethical hacking is a critical component of modern cybersecurity practices. It involves a structured methodology, ethical considerations, and a commitment to safeguarding digital assets. By proactively identifying and mitigating vulnerabilities, ethical hackers play a pivotal role in protecting organizations and individuals from cyber threats. As technology continues to advance, the role of ethical hackers will remain indispensable in maintaining the security and integrity of digital ecosystems.

In the ever-evolving landscape of cybersecurity, staying ahead of potential threats is paramount. While ethical hacking focuses on securing systems and networks, offensive security takes a different approach. Offensive security, often referred to as penetration testing or simply "pen testing," involves simulating cyberattacks to identify vulnerabilities and weaknesses in an organization's defenses. Python, with its versatility and extensive library support, has become a prominent choice for offensive security professionals. In this chapter, we'll explore how Python is employed in offensive security, the tools and techniques used, and the ethical considerations that guide these activities.

The Role of Offensive Security:

Offensive security, unlike ethical hacking, involves intentionally probing systems and networks with the goal of exploiting vulnerabilities. Organizations employ offensive security professionals to assess their cybersecurity posture, discover weaknesses, and remediate issues before malicious hackers can exploit them. By conducting controlled attacks, offensive security experts help organizations identify and close security gaps, ultimately fortifying their defenses.

Python as the Go-To Language:

Python's popularity in offensive security can be attributed to its flexibility, simplicity, and extensive library ecosystem. Offensive security professionals often use Python for a range of tasks, from developing custom exploits to automating attacks and managing the various phases of penetration testing. The language's readability and versatility make it an ideal choice for scripting and tool development in this field.

Common Python Tools in Offensive Security:

Several Python tools and frameworks are widely used in offensive security activities:

Metasploit Framework: Although primarily developed in Ruby, Metasploit also includes modules and scripts written in Python. It is a powerful platform for developing, testing, and executing exploits against vulnerable systems.

Scapy: Scapy is a Python library that allows for the creation, manipulation, and sending of network packets. Offensive security professionals use it to craft custom network packets for various attacks, such as network reconnaissance and packet sniffing.

PyCryptodome: Cryptography is essential in offensive security, especially when dealing with secure communications and encryption. PyCryptodome provides a wide range of cryptographic functions for Python applications.

Impacket: Impacket is a Python library that facilitates network communication and protocol exploitation. It is particularly valuable for attacking Windows systems and protocols.

PowerSploit: PowerSploit is a collection of PowerShell scripts that offensive security professionals often use to perform Windows-based attacks. While PowerShell is a Windows-native scripting language, Python can be used to automate the execution of these scripts.

The Offensive Security Process:

Offensive security follows a structured process similar to ethical hacking. This process includes several key phases:

Information Gathering: Just as in ethical hacking, offensive security begins with information gathering. Professionals collect data about the target, such as IP addresses, domain names, and employee details. Open-source intelligence (OSINT) plays a crucial role in this phase.

Scanning and Enumeration: Offensive security experts employ various scanning tools and techniques to identify potential entry points and vulnerabilities. Enumeration involves extracting detailed information about the target's systems and services.

Exploitation: This phase focuses on leveraging vulnerabilities to gain unauthorized access. Offensive security professionals use Python scripts to automate attacks, exploit vulnerabilities, and maintain control over compromised systems.

Post-Exploitation: After gaining access, offensive security experts aim to maintain their presence, expand their control, and gather sensitive information. Python scripts are instrumental in performing these tasks efficiently.

Reporting: As with ethical hacking, the final phase involves reporting the findings to the organization. However, in offensive security, the emphasis is on providing detailed information about vulnerabilities exploited, the potential impact, and recommended mitigations.

Ethical Considerations in Offensive Security:

While offensive security involves controlled attacks, ethical considerations remain paramount. Offensive security professionals must adhere to strict ethical guidelines to ensure that their actions are legal, responsible, and authorized by the organization. Obtaining proper consent and authorization is critical, as is ensuring that any actions taken do not result in harm or disruption to the target organization.

The Evolution of Offensive Security:

As cyber threats continue to evolve, offensive security practices must adapt accordingly. Python's role in offensive security is likely to grow, given its adaptability and the ever-expanding library ecosystem. Offensive security professionals will continue to rely on Python as they develop new tools and techniques to counter emerging threats.

In summary, offensive security is a crucial aspect of modern cybersecurity practices. Python's versatility and extensive library support make it an invaluable tool in the arsenal of offensive security professionals. As organizations recognize the importance of proactively identifying vulnerabilities and weaknesses in their digital defenses, the role of Python in offensive security will only become more significant. With ethical considerations at the forefront, offensive security professionals play a pivotal role in helping organizations stay ahead of potential cyber threats and secure their digital assets.

Chapter 5: Natural Language Processing and AI

Natural Language Processing (NLP) is a fascinating field that intersects computer science, artificial intelligence, and linguistics. It's all about enabling machines to understand, interpret, and generate human language. NLP has seen tremendous growth in recent years, with practical applications ranging from chatbots and virtual assistants to sentiment analysis and language translation. In this chapter, we'll embark on a journey to explore the world of NLP with Python, a versatile and widely-used programming language in the field of text and language processing.

Why NLP Matters:

Language is one of the primary means through which humans communicate, share information, and express themselves. NLP seeks to bridge the gap between human language and machine understanding. It opens up a wealth of possibilities for automating text-related tasks and extracting valuable insights from vast amounts of textual data.

Python: The Go-To Language for NLP:

Python's popularity in NLP can be attributed to several key factors. First, it boasts a large and active community of developers and researchers in the field. This community has contributed numerous libraries and tools specifically designed for NLP tasks. Second, Python's syntax is clear and readable, making it an excellent choice for developing NLP applications. Finally, Python's compatibility with various data analysis and machine learning libraries like NumPy, pandas, scikit-learn, and TensorFlow makes it a natural fit for NLP projects.

Fundamental NLP Tasks:

Before diving into Python-specific NLP tools, it's essential to understand the fundamental tasks in the field. These tasks include:

Tokenization: Breaking text into smaller units, often words or sentences, is called tokenization. Python provides libraries like NLTK and spaCy for this purpose.

Part-of-Speech Tagging (POS): POS tagging assigns grammatical categories (e.g., noun, verb, adjective) to each word in a sentence. Libraries like NLTK and spaCy excel in this task.

Named Entity Recognition (NER): NER identifies and classifies entities like names of people, places, and organizations in text. spaCy and NLTK offer NER capabilities.

Text Classification: Text classification involves assigning predefined labels or categories to text documents. Python's scikit-learn and TensorFlow are commonly used for this task.

Sentiment Analysis: Sentiment analysis determines the emotional tone or sentiment expressed in a piece of text. Python libraries such as TextBlob and VADER are popular choices for sentiment analysis.

Language Translation: Language translation, exemplified by Google Translate, uses machine learning models for translating text from one language to another.

Python Libraries for NLP:

Python provides a rich ecosystem of libraries and frameworks tailored for NLP tasks:

Natural Language Toolkit (NLTK): NLTK is a comprehensive library that covers a wide range of NLP tasks. It includes various datasets, corpora, and pre-trained models for NLP research.

spaCy: spaCy is a fast and efficient library designed for NLP professionals. It offers tokenization, POS tagging, NER, and dependency parsing, among other features.

TextBlob: TextBlob simplifies NLP tasks with a user-friendly interface. It's built on NLTK and Pattern and provides easy-to-use methods for tasks like sentiment analysis and translation.

Gensim: Gensim specializes in topic modeling and document similarity analysis. It's particularly useful for tasks like document clustering and creating word embeddings.

Transformers: The Hugging Face Transformers library has revolutionized the NLP landscape with pre-trained models like BERT, GPT-2, and RoBERTa. It makes it easier than ever to perform tasks like text generation, text classification, and more.

Practical NLP Projects:

To get a hands-on experience with NLP in Python, you can start with practical projects:

Sentiment Analysis for Social Media: Build a sentiment analysis tool that evaluates the sentiment of tweets or Facebook posts. Use TextBlob or VADER for this task.

Chatbot Development: Create a chatbot that can interact with users and respond to their queries. Libraries like NLTK and spaCy can help with text processing.

Document Summarization: Develop a tool that summarizes lengthy documents or articles. This project can involve extracting key sentences or phrases to condense the content.

Language Translation: Build a language translation app using pre-trained translation models from the Hugging Face Transformers library.

Named Entity Recognition (NER): Create an NER model that can identify and classify entities in a given text, such as names of people, places, and organizations.

Ethical Considerations:

When working with NLP, it's crucial to consider ethical implications. NLP models can inadvertently reinforce biases present in training data, leading to unfair or discriminatory outcomes. It's essential to address bias, fairness, and privacy concerns in NLP applications. Also, obtaining proper consent for data collection and usage is crucial, especially when dealing with user-generated text.

Conclusion:

NLP with Python opens up exciting possibilities for automating language-related tasks, extracting insights from textual data, and enhancing human-computer interaction. Python's extensive library support and ease of use make it an ideal choice for NLP practitioners. Whether you're interested in sentiment analysis, language translation, chatbots, or any other NLP application, Python provides the tools and resources to get started on your journey into the world of natural language processing.

Artificial Intelligence (AI) and Machine Learning (ML) are rapidly evolving fields that have the potential to transform various industries, from healthcare and finance to transportation and entertainment. In this chapter, we'll embark on a journey to explore the implementation of AI and ML algorithms using Python, a versatile and widely-used programming language in these domains. We'll delve into the foundational concepts, popular libraries, and practical applications that make AI and ML so exciting and relevant in today's world.

Why AI and ML Matter:

AI and ML are driving innovation across many domains by enabling computers to learn from data, make predictions, and perform tasks that traditionally required human intelligence. These technologies can process vast amounts of data, uncover patterns, and generate insights that can lead to more informed decision-making and automation of complex processes.

Python: The Go-To Language for AI and ML:

Python has emerged as the dominant programming language for AI and ML for several reasons. First, it boasts an extensive ecosystem of libraries and frameworks dedicated to these fields, such as TensorFlow, PyTorch, and scikit-learn. Second, Python's simplicity and readability make it accessible to both beginners and experts. Third, it provides excellent support for data manipulation and visualization, which are essential in AI and ML workflows. These factors have made Python the preferred language for implementing AI and ML algorithms.

Key Concepts in AI and ML:

Before diving into Python-specific implementations, it's essential to grasp the core concepts in AI and ML:

Supervised Learning: In supervised learning, the algorithm learns from a labeled dataset, where each input is associated with the correct output. Common tasks include classification and regression.

Unsupervised Learning: Unsupervised learning involves learning patterns and structures from unlabeled data. Clustering and dimensionality reduction are typical unsupervised tasks.

Deep Learning: Deep learning is a subset of ML that focuses on neural networks with many layers (deep neural networks). It excels in tasks like image recognition, natural language processing, and speech recognition.

Reinforcement Learning: Reinforcement learning is about training agents to make sequential decisions in an environment to maximize a reward. It's often used in robotics and game playing.

Python Libraries for AI and ML:

Python's rich ecosystem includes a plethora of libraries and frameworks tailored for AI and ML tasks:

TensorFlow: Developed by Google, TensorFlow is one of the most popular deep learning frameworks. It offers high-level APIs like Keras and lower-level APIs for flexibility.

PyTorch: PyTorch is known for its dynamic computation graph and is favored by researchers and developers for its flexibility and ease of use.

scikit-learn: scikit-learn is a versatile library for classical ML tasks. It provides tools for data preprocessing, model selection, and evaluation.

Pandas: Pandas is a powerful library for data manipulation and analysis. It's invaluable for data preprocessing tasks before feeding data into ML models.

Matplotlib and Seaborn: These libraries are essential for data visualization, helping you understand your data and model results.

Practical AI and ML Projects:

To gain hands-on experience with AI and ML in Python, consider starting with practical projects:

Image Classification: Implement an image classification model using a pre-trained deep learning model (e.g., VGG16 or ResNet) and fine-tune it for a specific task.

Natural Language Processing: Build a sentiment analysis model using a recurrent neural network (RNN) or a transformer-based model like BERT.

Recommender System: Create a movie or product recommendation system using collaborative filtering techniques or matrix factorization.

Anomaly Detection: Develop an anomaly detection system for fraud detection or network security using unsupervised learning methods.

Chatbot: Build a chatbot that can interact with users, answer questions, and perform tasks using natural language understanding.

Ethical Considerations:

Ethical considerations are paramount in AI and ML implementations. Bias in training data, fairness, transparency, and privacy are critical issues to address. It's essential to assess the potential biases in your data and models and take steps to mitigate them. Additionally, AI systems should be designed with user consent and data security in mind.

Conclusion:

AI and ML are revolutionizing industries by automating tasks, making predictions, and driving data-driven decision-making. Python's prominence in these fields is a testament to its versatility and the wealth of libraries available to developers. Whether you're interested in image recognition, natural language understanding, recommendation systems, or anomaly detection, Python provides the tools and resources to bring your AI and ML projects to life. Remember to approach these technologies with ethical considerations in mind, ensuring that your implementations prioritize fairness, transparency, and user privacy.

Chapter 6: IoT Integration and Automation

The Internet of Things, or IoT, is a transformative technology that has the potential to change the way we interact with the physical world around us. It involves connecting everyday objects, devices, and sensors to the internet, allowing them to collect and exchange data. Python, with its versatility and wide range of libraries, is an excellent choice for developing applications to connect and control IoT devices. In this chapter, we'll explore how Python can be used to connect and communicate with IoT devices, empowering you to build innovative and intelligent systems.

Understanding the Internet of Things (IoT):

IoT encompasses a vast array of devices, from smart thermostats and wearable fitness trackers to industrial sensors and autonomous vehicles. These devices are equipped with sensors, actuators, and communication capabilities, enabling them to gather data, process information, and take actions based on that data. The goal of IoT is to enhance efficiency, improve decision-making, and create more connected and responsive environments.

Python's Role in IoT Development:

Python has gained significant traction in the IoT landscape due to several key advantages. First and foremost, Python is easy to learn and use, making it accessible to a broad audience, including software developers, data scientists, and hobbyists. Its simplicity and readability enable developers to quickly prototype and develop IoT solutions.

Second, Python has a thriving ecosystem of libraries and frameworks that facilitate IoT development. Libraries like **MicroPython** and **CircuitPython** are lightweight implementations of Python specifically designed for microcontrollers and small IoT devices. These implementations make it possible to write Python code that runs directly on IoT hardware, simplifying the development process.

Third, Python offers excellent support for data manipulation and analysis, which is crucial for processing the data generated by IoT devices. Libraries like **pandas** and **NumPy** are invaluable for managing and analyzing large datasets.

Types of IoT Devices:

IoT devices come in various forms and serve diverse purposes. Here are some common types:

Sensors: These devices capture data from the physical world, such as temperature, humidity, light, or motion. Examples include environmental sensors and accelerometers.

Actuators: Actuators are devices that can perform actions based on data or commands received. Examples include motors, servos, and relays.

Microcontrollers: Microcontrollers are small computing devices with limited processing power and memory. They are the brains of many IoT devices and often run Python.

Single-Board Computers (SBCs): SBCs like the Raspberry Pi and BeagleBone are more powerful than microcontrollers and can run full-fledged operating systems, including Python.

Connecting to IoT Devices:

Connecting to IoT devices typically involves the following steps:

Device Discovery: Discovering IoT devices on a network is the first step. This can be done using protocols like Simple Service Discovery Protocol (SSDP) or mDNS.

Communication Protocols: IoT devices use various communication protocols to exchange data. Common protocols include HTTP/HTTPS, MQTT (Message Queuing Telemetry Transport), CoAP (Constrained Application

Protocol), and WebSocket. Python libraries such as **requests,** **paho-mqtt**, and **websockets** simplify communication over these protocols.

Authentication and Security: Ensuring the security of IoT communication is critical. Using secure authentication mechanisms like OAuth, API keys, or SSL/TLS encryption helps protect sensitive data.

Data Parsing: IoT devices transmit data in various formats, including JSON, XML, or custom binary formats. Python's built-in libraries for parsing and processing data, such as **json** and **xml.etree.ElementTree,** make it straightforward to work with different data formats.

IoT Applications:

The applications of IoT are extensive and span across various industries. Some notable examples include:

Smart Homes: IoT devices like smart thermostats, lighting systems, and voice assistants enable homeowners to control and monitor their homes remotely.

Healthcare: Wearable IoT devices, such as fitness trackers and medical sensors, help individuals monitor their health and provide valuable data to healthcare professionals.

Agriculture: IoT sensors can monitor soil conditions, weather, and crop health, enabling farmers to optimize crop yields and reduce resource usage.

Manufacturing: IoT devices in manufacturing plants collect real-time data on equipment performance, enabling predictive maintenance and reducing downtime.

Smart Cities: IoT is used in traffic management, waste collection, and environmental monitoring to make cities more efficient and sustainable.

Challenges in IoT Development:

While IoT offers immense potential, it also presents challenges, including:

Security: IoT devices are often vulnerable to cyberattacks. Developers must prioritize security measures to protect data and privacy.

Interoperability: IoT devices from different manufacturers may use incompatible communication protocols, making interoperability a challenge.

Scalability: Handling large volumes of data generated by IoT devices requires robust infrastructure and data management.

Power Efficiency: Many IoT devices run on batteries, so power efficiency is crucial to extend battery life.

Conclusion:

The Internet of Things is reshaping our world by connecting everyday objects to the internet and enabling them to collect and exchange data. Python, with its accessibility, extensive libraries, and data processing capabilities, is a powerful tool for developing IoT solutions. Whether you're interested in building a smart home, monitoring environmental conditions, or optimizing industrial processes, Python can help you connect, communicate with, and control IoT devices effectively. As you embark on your IoT journey, remember to consider security, interoperability, scalability, and power efficiency to create robust and reliable IoT systems.

The Internet of Things (IoT) has ushered in a new era of automation and connectivity, revolutionizing the way we interact with the physical world. In this chapter, we'll delve into the exciting world of building IoT solutions for automation using Python. As we embark on this journey, you'll discover how Python empowers developers to create intelligent and automated systems that can make our lives more convenient and efficient.

Understanding the Power of IoT Automation:

At its core, IoT automation involves using interconnected devices and sensors to collect data and perform actions without direct human intervention. This automation can range from simple tasks like turning on lights when motion is detected to complex processes like optimizing industrial production lines based on real-time data.

The Role of Python in IoT Automation:

Python is a versatile and user-friendly programming language that has gained significant popularity in the IoT domain. Its simplicity and readability make it an excellent choice for developing automation solutions. Python's rich ecosystem of libraries and frameworks provides developers with the tools they need to connect, control, and automate a wide range of IoT devices.

Components of an IoT Automation System:

An IoT automation system typically consists of the following components:

IoT Devices: These are the physical sensors, actuators, and devices that collect data and perform actions. Examples Include temperature sensors, smart locks, and robotic arms.

Communication Protocols: IoT devices communicate with each other and with central systems using various protocols. Common protocols include MQTT, CoAP, and HTTP, and Python provides libraries to work with these protocols efficiently.

Data Processing and Analysis: Collected data needs to be processed and analyzed to derive meaningful insights. Python's data analysis libraries, such as pandas and NumPy, come in handy for this purpose.

Cloud or Edge Computing: Depending on the complexity of the automation, data processing can occur either in the cloud or at the edge. Cloud platforms like AWS, Azure, and Google Cloud provide robust IoT services, while edge computing devices like the Raspberry Pi can perform local data processing.

User Interfaces: Many IoT automation systems require user interfaces for monitoring and control. Python's frameworks like Flask and Django can be used to develop web-based interfaces for interacting with IoT devices.

Developing IoT Automation Solutions with Python:

Here's a step-by-step guide to building IoT automation solutions with Python:

1. Identify Your Use Case: Begin by identifying the specific task or problem you want to automate with IoT. Whether it's home automation, industrial optimization, or environmental monitoring, a clear use case will guide your development efforts.

2. Choose the Right Hardware: Depending on your use case, select the appropriate IoT devices and sensors. Ensure they are compatible with Python or have libraries and drivers available for Python development.

3. Connect and Collect Data: Use Python libraries like MQTT, CoAP, or HTTP to establish connections with your IoT devices. Collect data from sensors and devices and store it for analysis.

4. Data Analysis: Use Python's data analysis libraries to process and analyze the collected data. You can identify patterns, anomalies, or trends that will inform your automation logic.

5. Implement Automation Logic: Based on your data analysis, develop automation logic using Python. This can involve setting up rules, triggers, and actions that dictate how IoT devices respond to specific conditions.

6. Implement User Interfaces: If your automation system requires user interaction, create a user interface using Python web frameworks like Flask or Django. This interface can provide real-time data visualization and control options.

7. Test and Iterate: Thoroughly test your IoT automation solution in a controlled environment. Identify any issues or performance bottlenecks and refine your code as needed.

8. Deployment: Deploy your solution in the target environment, whether it's your home, a factory floor, or a smart city. Monitor the system's performance and make adjustments as necessary.

IoT Automation Use Cases:

IoT automation has a wide range of practical use cases across various domains:

1. Smart Homes: Automate lighting, heating, security, and entertainment systems for enhanced comfort and energy efficiency.

2. Industrial Automation: Optimize manufacturing processes, monitor equipment health, and reduce downtime in factories.

3. Environmental Monitoring: Use IoT sensors to collect data on air quality, weather conditions, and pollution levels.

4. Agriculture: Implement automation for precision farming, irrigation control, and livestock monitoring.

5. Healthcare: Develop wearable IoT devices for remote patient monitoring and medication adherence.

Challenges in IoT Automation:

While IoT automation holds tremendous promise, it also presents challenges:

1. Security: Ensuring the security of IoT devices and data is crucial, as vulnerabilities can lead to breaches and privacy issues.

2. Scalability: As the number of connected devices grows, managing and scaling an IoT ecosystem can become complex.

3. Interoperability: Different IoT devices may use incompatible communication protocols, necessitating careful integration.

4. Power Efficiency: Many IoT devices are battery-powered, so optimizing power consumption is essential for extended battery life.

Conclusion:

IoT automation powered by Python is transforming the way we interact with our environment and simplifying our lives. From smart homes to industrial optimization, the possibilities are endless. Python's simplicity, vast library ecosystem, and data analysis capabilities make it a perfect choice for building IoT automation solutions that can bring convenience, efficiency, and intelligence to our daily lives. As you explore the world of IoT automation, remember that thoughtful planning, robust development, and ongoing maintenance are key to success in this exciting field.

Chapter 7: Data Engineering and Big Data Processing

Exploring the vast landscape of Big Data can be both exciting and daunting, my friend. In today's data-driven world, businesses and organizations are collecting massive amounts of data at an unprecedented rate. But what exactly is Big Data, and why does it matter? Let's embark on a journey to unravel the concepts and the ecosystem that underpin this fascinating realm.

At its core, Big Data refers to the immense volume, variety, and velocity of data generated daily across digital platforms and devices. This data comes from a multitude of sources, including social media interactions, online transactions, sensor data, and much more. The sheer volume is staggering, often measured in petabytes or even exabytes, and it continues to grow exponentially.

One of the defining characteristics of Big Data is its variety. Data comes in various formats and types, including structured, semi-structured, and unstructured data. Structured data is highly organized and easy to query, such as information stored in relational databases. Semi-structured data, like XML or JSON, has some structure but doesn't fit neatly into tables. Unstructured data, on the other hand, lacks a predefined structure and can include text documents, images, audio, and video files.

Velocity, another crucial aspect, refers to the speed at which data is generated and needs to be processed. In the realm of Big Data, real-time or near-real-time processing is often required to derive valuable insights and make timely decisions. Think of financial transactions, where milliseconds can make a significant difference.

To make sense of this deluge of data, organizations turn to Big Data technologies and tools. The Big Data ecosystem comprises a multitude of components, and let's dive into some of the key ones.

Storage: Big Data storage solutions like Hadoop Distributed File System (HDFS) and cloud-based storage platforms provide the capacity to store vast datasets. These systems distribute data across multiple nodes for redundancy and scalability.

Processing: Big Data processing frameworks, such as Apache Hadoop and Apache Spark, enable the parallel processing of data across distributed clusters of computers. This approach significantly reduces processing time for large datasets.

Data Ingestion: Data must be ingested from various sources into the Big Data ecosystem. Tools like Apache Kafka and Apache Nifi handle data ingestion, ensuring data flows smoothly into storage and processing engines.

Databases: NoSQL databases, like MongoDB and Cassandra, have gained popularity in Big Data applications. They can handle unstructured and semi-structured data efficiently, making them suitable for a variety of use cases.

Analytics and Machine Learning: Tools and libraries for advanced analytics and machine learning, such as Apache Flink and TensorFlow, allow organizations to extract valuable insights from their data. Machine learning models can identify patterns and make predictions.

Data Visualization: Communicating insights effectively is crucial, and data visualization tools like Tableau and Power BI help turn raw data into meaningful visuals that support decision-making.

Security and Compliance: With the growing concern around data privacy and security, Big Data ecosystems include components for securing data, managing access control, and ensuring compliance with regulations like GDPR and HIPAA.

Now, you might wonder why Big Data matters and how it can benefit various industries. The answers are as diverse as the data itself. Let's explore a few examples:

1. Healthcare: Big Data analytics can help healthcare providers improve patient care by analyzing electronic health records, monitoring patient vital signs in real-time, and predicting disease outbreaks.

2. Retail: Retailers use Big Data to personalize customer experiences, optimize supply chains, and forecast demand more accurately, leading to reduced costs and increased sales.

3. Finance: Financial institutions rely on Big Data for fraud detection, risk assessment, and algorithmic trading, where split-second decisions can make or break investments.

4. Manufacturing: Manufacturers leverage Big Data to enhance production efficiency, monitor equipment health with IoT sensors, and predict maintenance needs, reducing downtime and costs.

5. Transportation: In the transportation sector, Big Data helps improve route optimization, traffic management, and vehicle maintenance, leading to smoother commutes and reduced environmental impact.

Challenges in the World of Big Data:

While Big Data offers immense possibilities, it also presents significant challenges. One of the most significant hurdles is data quality. With such a vast volume of data from various sources, ensuring data accuracy, completeness, and consistency is a considerable challenge.

Another challenge is data privacy and security. As organizations collect and analyze more data, there's an increasing risk of data breaches and privacy violations. Striking a balance between data access and security is crucial.

Scalability and infrastructure complexity are additional challenges. Building and maintaining a Big Data ecosystem can be costly and complex, requiring skilled professionals to manage and optimize.

The Future of Big Data:

The Big Data landscape continues to evolve rapidly. Trends like edge computing, which brings data processing closer to the data source, and federated learning, which preserves data privacy while enabling machine learning on decentralized data, are gaining traction.

Certainly, let's delve into the fascinating world of Python for data engineering and processing. Python has established itself as a versatile and powerful language for handling data at various scales, from small datasets to massive Big Data pipelines. In this chapter, we'll explore the key concepts, libraries, and techniques that make Python an invaluable tool for data engineers.

At its core, data engineering involves the collection, transformation, and storage of data for further analysis and decision-making. Python's simplicity and a rich ecosystem of libraries make it well-suited for these tasks. Let's start by examining the primary steps in the data engineering process.

Data Ingestion: The journey begins with data ingestion, where data is collected from various sources, such as databases, APIs, log files, and sensors. Python offers libraries like **requests** for web scraping and APIs, **pandas** for reading structured data, and tools like **BeautifulSoup** for parsing HTML.

Data Transformation: Once the data is ingested, it often needs to be cleaned, transformed, and structured for analysis. Python's **pandas** library is a powerhouse for data manipulation. It allows you to filter, aggregate, pivot, and reshape data with ease.

Data Storage: Data engineers need reliable storage solutions to store and manage datasets. Python can interact with a wide range of databases, both relational (e.g., MySQL, PostgreSQL) and NoSQL (e.g., MongoDB, Cassandra), using libraries like **SQLAlchemy** and **pymongo**.

Data Pipelines: For larger-scale data processing, data engineers often build data pipelines. These pipelines automate the flow of data, ensuring that data is ingested, transformed, and stored efficiently. Libraries like **Apache Airflow** are commonly used for pipeline orchestration.

Distributed Computing: Handling Big Data requires distributed computing frameworks, and Python offers robust tools for this purpose. Libraries like **Dask** and **PySpark** allow you to scale your data processing tasks across clusters of machines.

Real-Time Data Processing: Some use cases demand real-time data processing. Python shines in this area with libraries like **Kafka** for stream processing and **Redis** for in-memory data storage.

Now, let's take a closer look at some of the key Python libraries that data engineers rely on:

1. Pandas: Pandas is the go-to library for data manipulation and analysis. It provides data structures like DataFrames and Series, making it easy to filter, aggregate, and transform data.

2. NumPy: NumPy is a fundamental library for scientific computing in Python. It provides support for arrays and matrices, which are essential for numerical operations and data handling.

3. SQLAlchemy: SQLAlchemy is a powerful and flexible library for working with relational databases. It offers an Object-Relational Mapping (ORM) system and enables you to interact with databases using Python objects.

4. Apache Spark: While not entirely written in Python, PySpark is a Python API for Apache Spark, a distributed data processing framework. It's perfect for processing large datasets in parallel.

5. Dask: Dask is a parallel computing library that extends Python's capabilities to larger-than-memory, multi-core, and distributed computing. It's excellent for handling Big Data workloads.

6. Apache Kafka: Kafka is a distributed streaming platform that allows you to build real-time data pipelines and streaming applications. Python has libraries like **confluent-kafka** for Kafka integration.

7. Redis: Redis is an in-memory data store that is often used for caching and real-time data processing. Python provides a **redis** library for interacting with Redis.

8. Apache Airflow: Apache Airflow is an open-source platform for orchestrating complex data workflows. You can define, schedule, and monitor data pipelines using Python code.

Use Cases for Python in Data Engineering:

Python's versatility and extensive ecosystem make it suitable for various data engineering use cases. Here are a few examples:

1. ETL (Extract, Transform, Load) Pipelines: Python is commonly used to build ETL pipelines that extract data from different sources, transform it into a structured format, and load it into a data warehouse or database.

2. Data Integration: Python can integrate data from multiple sources, whether they are APIs, databases, or flat files, and combine them into a unified dataset for analysis.

3. Data Quality and Validation: Python scripts can be employed to validate and clean data, ensuring its accuracy and consistency.

4. Real-Time Data Processing: Python, along with tools like Kafka, can process streaming data in real-time, enabling businesses to make instant decisions based on fresh data.

5. Batch Processing: For batch processing of large datasets, Python libraries like Dask and PySpark help distribute computation across multiple machines.

Challenges in Data Engineering with Python:

While Python offers numerous advantages, data engineers also face challenges when working with the language. One common challenge is performance, particularly when dealing with very large datasets. In such cases, optimizing code and leveraging distributed computing frameworks become crucial.

Data engineers must also consider data security and privacy, especially when dealing with sensitive information. Proper encryption and access controls are essential to protect data.

In summary, Python's role in data engineering is ever-expanding, thanks to its user-friendly syntax and a vast ecosystem of libraries and tools. Whether you're building data pipelines, processing real-time streams, or managing databases, Python provides a flexible and powerful foundation for data engineering tasks.

Chapter 8: Advanced Cloud and Container Orchestration

Absolutely, let's embark on an exciting journey into the realm of cloud orchestration with Python. Cloud orchestration is a pivotal aspect of modern cloud computing, enabling the efficient management, provisioning, and automation of cloud resources. In this chapter, we will explore how Python, with its versatility and extensive library support, plays a vital role in orchestrating cloud environments.

Understanding Cloud Orchestration:

Before we dive into Python's role in cloud orchestration, let's establish a clear understanding of what cloud orchestration entails. Cloud orchestration is the process of automating and coordinating tasks related to cloud infrastructure. It involves deploying, scaling, managing, and monitoring cloud resources to ensure they meet the demands of your applications and services.

Python as a Glue Language:

Python is often referred to as a "glue language" due to its ability to seamlessly integrate and orchestrate various components and services. When it comes to cloud orchestration, Python shines as a glue language that binds together different cloud providers, services, and infrastructure components.

Infrastructure as Code (IaC):

One of the fundamental concepts in cloud orchestration is Infrastructure as Code (IaC). IaC allows you to define and manage cloud infrastructure using code. Python excels in this area with libraries like Terraform, AWS Cloud Development Kit (CDK), and Ansible. These tools enable you to describe your cloud resources in code, making it easy to provision and manage infrastructure.

Cloud Providers' SDKs:

Python provides Software Development Kits (SDKs) for major cloud providers, such as Amazon Web Services (AWS), Microsoft Azure, Google Cloud Platform (GCP), and more. These SDKs empower developers to interact with cloud services programmatically. For instance, **boto3** is the AWS SDK for Python, allowing you to manage AWS resources through Python code.

Container Orchestration:

Containers have revolutionized application deployment and orchestration. Python is instrumental in managing container orchestration platforms like Kubernetes. The **kubectl** Python library enables you to interact with Kubernetes clusters and automate tasks like deploying and scaling containerized applications.

Serverless Computing:

Serverless computing abstracts infrastructure management, enabling developers to focus solely on code. Python is a popular choice for building serverless functions on platforms like AWS Lambda and Azure Functions. You can write Lambda functions in Python, responding to events and triggers.

Multi-Cloud Orchestration:

Many organizations operate in multi-cloud environments, using multiple cloud providers to diversify their infrastructure or reduce vendor lock-in. Python's portability makes it an ideal choice for orchestrating resources across various cloud providers, ensuring consistent deployment and management.

DevOps and Continuous Integration/Continuous Deployment (CI/CD):

Python plays a crucial role in DevOps practices and CI/CD pipelines. Tools like Jenkins, Travis CI, and CircleCI can be configured to automate the deployment of applications and services on the cloud. Python scripts can interact with these tools, allowing for streamlined, automated deployments.

Monitoring and Logging:

Effective orchestration involves monitoring the health and performance of cloud resources. Python can be employed to create custom monitoring and logging solutions using libraries like Prometheus and Grafana. Additionally, Python's rich ecosystem includes logging frameworks like **loguru** and **structlog** for comprehensive logging.

Challenges in Cloud Orchestration:

While Python is an excellent choice for cloud orchestration, it's essential to be aware of potential challenges. Some of these challenges include:

Complexity: Orchestrating cloud resources can become complex as the scale and diversity of services grow. Python's flexibility can mitigate this, but proper architecture and design are crucial.

Security: Managing access controls, encryption, and authentication in a cloud environment is paramount. Python scripts handling security-sensitive operations must be thoroughly secured to prevent vulnerabilities.

Cost Management: Orchestrating resources efficiently helps control costs. Python scripts can optimize resource allocation, but careful monitoring and management are essential to avoid unexpected expenses.

Error Handling: Handling errors and exceptions in cloud orchestration scripts is critical. Python offers robust error-handling capabilities, but thorough testing and monitoring are essential.

In summary, Python is a powerful ally in the world of cloud orchestration. Its versatility, extensive libraries, and vibrant community support make it an ideal choice for managing, automating, and orchestrating cloud infrastructure and services. As organizations increasingly embrace cloud computing, Python's role in cloud orchestration will continue to grow, empowering engineers and developers to streamline and automate complex cloud workflows.

Let's delve into the fascinating world of containerization and orchestration tools, which have become indispensable in modern software development and deployment. These technologies address the challenges of managing complex applications, ensuring scalability, and improving resource utilization. In this chapter, we'll explore containerization, container orchestration, and some popular tools that enable efficient management of containers.

Understanding Containerization:

Containerization is a method of packaging, distributing, and running applications and their dependencies in isolated, lightweight containers. Containers are portable, consistent, and can run on various environments, making them an ideal solution for modern application deployment. Docker, the pioneering containerization platform, played a significant role in popularizing this technology.

The Role of Docker:

Docker, often synonymous with containerization, provides an ecosystem for building, sharing, and running containers. Docker containers encapsulate applications, libraries, and dependencies, ensuring consistent behavior across different environments. Developers can create, test, and deploy containers effortlessly, making Docker a go-to tool for many.

Container Images and Registries:

Containerization begins with creating container images. These images serve as blueprints for containers, defining the application and its environment. Docker Hub, a public registry, houses a vast collection of pre-built images, offering convenience for developers. Organizations often use private container registries like Amazon ECR or Google Container Registry to manage proprietary images securely.

Container Orchestration:

While containers solve the problem of packaging applications, container orchestration addresses the complexities of deploying, scaling, and managing containerized services. It automates tasks like load balancing, scaling, self-healing, and service discovery.

Introduction to Kubernetes:

Kubernetes, often abbreviated as K8s, is the leading container orchestration platform. It was originally developed by Google and is now maintained by the Cloud Native Computing Foundation (CNCF). Kubernetes abstracts the underlying infrastructure, providing a unified API to manage containers across on-premises, cloud, and hybrid environments.

Kubernetes Components:

Kubernetes comprises a set of components that work together to orchestrate containers. The control plane includes the Kubernetes API server, etcd for configuration storage, and various controllers. Nodes in the cluster run the container runtime (like Docker or containerd) and the kubelet, which manages containers on the node.

Container Orchestration Features:

Kubernetes offers powerful features for container orchestration:

Scalability: Kubernetes can scale applications horizontally or vertically based on demand. It automatically manages the allocation of containers to nodes.

Self-Healing: If a container or node fails, Kubernetes reschedules containers to healthy nodes, ensuring high availability.

Service Discovery: Kubernetes provides DNS-based service discovery, simplifying communication between services.

Load Balancing: Built-in load balancers distribute traffic among containers, ensuring efficient resource utilization.

Rolling Updates: Kubernetes supports rolling updates, allowing for zero-downtime deployments.

Configuration Management: Kubernetes enables you to define and manage configuration settings separately from application code.

Alternatives to Kubernetes:

While Kubernetes is the dominant player, other container orchestration tools like Docker Swarm and Apache Mesos have their niches. Docker Swarm is known for its simplicity and integrates seamlessly with Docker, making it a good choice for smaller deployments. Apache Mesos is a general-purpose cluster manager that can orchestrate containers alongside other workloads.

Container Orchestration Challenges:

Despite the advantages, container orchestration comes with its challenges:

Complexity: Managing Kubernetes clusters can be complex, especially for small teams or organizations new to containerization.

Learning Curve: Teams need to invest time in learning Kubernetes concepts and best practices.

Resource Consumption: Kubernetes requires resources for its control plane, which might be overkill for smaller applications.

Infrastructure Choice: Deciding between managed Kubernetes services (like AWS EKS, GCP GKE) or self-hosting requires careful consideration.

The Future of Containerization and Orchestration:
Containerization and orchestration have transformed how we build, deploy, and manage applications. As the technology evolves, we can expect improvements in areas like security, observability, and serverless computing. Concepts like GitOps, where configurations are versioned and stored in Git repositories, are gaining traction as well.

Choosing the Right Tool:
The choice of containerization and orchestration tool depends on your specific needs, team expertise, and infrastructure. For small projects, Docker Compose may suffice, while Kubernetes excels in managing complex microservices architectures at scale. Consider factors like community support, ecosystem, and integration with existing tools when making your decision.

Conclusion:
Containerization and orchestration have reshaped the software development landscape. They offer portability, scalability, and automation, driving the adoption of cloud-native practices. Whether you choose Kubernetes, Docker Swarm, or another tool, mastering containerization and orchestration is essential for modern software engineers and organizations looking to stay competitive in the ever-evolving tech landscape.

Chapter 9: Advanced Testing and Test Automation

In the ever-evolving world of software development, testing plays a pivotal role in ensuring the quality, reliability, and functionality of applications. As you've progressed through this book, you've gained a solid understanding of Python, automation, and various tools and libraries. Now, let's delve into the realm of advanced testing strategies, where we'll explore techniques and practices that go beyond the basics to help you create robust and thoroughly tested software.

Unit Testing:

By now, you're likely familiar with unit testing, where individual components or functions are tested in isolation. This practice remains fundamental to maintaining code quality and catching errors early in the development process. However, in advanced testing, unit tests become even more critical. It's essential not only to test various input scenarios but also to consider edge cases and boundary conditions that can potentially break your code.

Test-Driven Development (TDD):

Test-Driven Development is a powerful methodology that guides your coding process. Instead of writing code first and then tests, you write tests before you even start coding. This approach forces you to think deeply about the expected behavior of your functions and classes, resulting in more robust and well-documented code. TDD encourages small, incremental development cycles, making it easier to catch and fix bugs as they emerge.

Integration Testing:

While unit testing focuses on individual components, integration testing ensures that these components work together harmoniously. In advanced testing, integration tests become indispensable, especially in complex systems with numerous interconnected modules. These tests validate the interactions between different parts of your application, helping to identify integration issues and ensuring that data flows correctly.

Continuous Integration (CI) and Continuous Deployment (CD):
CI/CD pipelines are essential components of modern software development workflows. CI involves automating the building and testing of your code every time changes are pushed to a version control repository, ensuring that new code doesn't introduce regressions. CD extends CI by automating the deployment process, allowing you to release new features or bug fixes seamlessly. Tools like Jenkins, Travis CI, and CircleCI are commonly used for CI/CD.

End-to-End (E2E) Testing:
E2E testing simulates user interactions with your application to verify its functionality from start to finish. These tests ensure that all components work together correctly, including the user interface. Selenium, Puppeteer, and Cypress are popular tools for E2E testing. In advanced testing, you should focus on creating comprehensive E2E test suites to catch any issues that might not surface in unit or integration tests.

Load and Performance Testing:
As your application gains users, it's crucial to assess how it performs under different loads. Load testing checks how your application handles increased traffic, while performance testing examines response times and resource utilization. Tools like Apache JMeter and Gatling are used for load and performance testing. Advanced testing includes stress testing to determine the system's breaking point and capacity planning to ensure it can handle anticipated loads.

Security Testing:
Security is paramount in software development, and advanced testing should include security testing practices. Tools like OWASP ZAP and Burp Suite help identify vulnerabilities such as cross-site scripting (XSS) and SQL injection. Regular security audits, code reviews, and penetration testing should be part of

your testing strategy to ensure your application is protected against potential threats.

Mutation Testing:

Mutation testing is an advanced technique that involves making small, deliberate changes (mutations) to your code to determine if your tests can detect these changes. If a mutation goes undetected, it indicates a weakness in your test suite. While mutation testing is computationally intensive and not commonly used, it can provide valuable insights into the effectiveness of your tests.

A/B Testing:

A/B testing, also known as split testing, allows you to compare two versions of a web page or application to determine which one performs better. This technique is widely used in marketing but can also be applied to software development to make data-driven decisions about features, user interfaces, and algorithms. Tools like Optimizely and Google Optimize facilitate A/B testing.

Monitoring and Observability:

In advanced testing, monitoring and observability become part of your testing strategy. Monitoring involves tracking the health and performance of your application in real-time, while observability focuses on understanding how your application behaves in different situations. Tools like Prometheus, Grafana, and Elasticsearch with Kibana can help you gain insights into your application's behavior.

Test Environments and Containers:

Advanced testing often involves creating isolated test environments that mimic production as closely as possible. Docker containers are invaluable for replicating these environments consistently. You can use container orchestration tools like Kubernetes to manage test environments efficiently, ensuring that they remain consistent across development, testing, and production.

Conclusion:

In the ever-evolving landscape of software development, advanced testing strategies are critical for producing high-quality software. From unit testing and TDD to integration, E2E, and security testing, a comprehensive testing strategy ensures that your code is robust, reliable, and secure. Incorporating advanced testing practices into your development workflow will help you catch and address issues early, ultimately leading to better software and happier users.

Test automation is an indispensable aspect of modern software development, enabling teams to deliver high-quality applications efficiently and reliably. As you progress in your journey of mastering test automation, it's essential to embrace best practices and utilize robust testing frameworks to maximize the effectiveness of your testing efforts. In this chapter, we will explore the key principles of test automation best practices and delve into some popular testing frameworks that can simplify and streamline your testing processes.

1. Early Involvement in the Development Lifecycle:

One of the fundamental principles of test automation is to involve testing early in the software development lifecycle. Automated tests can catch issues at the earliest stages, preventing defects from propagating downstream and reducing the cost of fixing them. This aligns with the Agile and DevOps philosophies, where testing is integrated throughout the development process.

2. Selective Test Automation:

Not all tests need to be automated. It's crucial to use a risk-based approach to determine which tests are candidates for automation. High-priority and high-impact scenarios, such as critical functionality and regression tests, should be automated first. This ensures that essential areas of your application are thoroughly tested.

3. Maintainable and Readable Tests:

Maintainability and readability are key aspects of successful test automation. Your test scripts should be easy to understand

and maintain by anyone on your team. Use descriptive test and method names, meaningful comments, and clear documentation. Employ coding standards and design patterns to create clean, modular, and maintainable test code.

4. Test Data Management:

Effective test data management is essential for successful automation. Ensure that your tests can work with different sets of test data, including boundary cases and edge scenarios. Consider using data-driven testing approaches, where test cases are separated from test data, allowing for easy data variations without modifying the test scripts.

5. Continuous Integration (CI) and Continuous Delivery (CD):

Integrate your automated tests into your CI/CD pipeline to enable continuous testing. This ensures that tests are executed automatically whenever changes are made to the codebase. Popular CI/CD platforms like Jenkins, Travis CI, and CircleCI provide integrations with various testing frameworks.

6. Parallel and Distributed Testing:

To accelerate testing and reduce execution time, explore parallel and distributed testing. Parallel testing runs multiple test cases concurrently on multiple machines or threads, while distributed testing distributes test execution across various environments. This approach is particularly useful for large test suites.

7. Robust Error Handling:

Your test automation scripts should include robust error handling to handle unexpected situations gracefully. Implement proper exception handling and logging to capture details of test failures. This aids in diagnosing and debugging issues quickly.

8. Regular Test Maintenance:

Automation scripts require regular maintenance to stay relevant as the application evolves. As new features are added or existing ones change, update your tests accordingly.

Regularly review and refactor test code to keep it efficient and maintainable.

9. Continuous Learning:

Stay up-to-date with the latest developments in test automation tools and practices. Attend training sessions, webinars, and conferences, and participate in online forums and communities to exchange knowledge and learn from others in the field. Investing in continuous learning will help you enhance your automation skills.

Popular Testing Frameworks:

Now, let's explore some widely used testing frameworks that can aid in implementing test automation best practices:

1. Selenium:

Selenium is a popular open-source testing framework for automating web applications. It supports multiple programming languages, including Python, and provides a range of features for interacting with web elements, simulating user actions, and validating web page behavior.

2. PyTest:

PyTest is a powerful and extensible testing framework for Python. It's known for its simplicity and readability, making it a favorite among Python developers. PyTest supports a wide range of plugins and can be easily integrated with Selenium and other testing tools.

3. Robot Framework:

Robot Framework is a versatile, keyword-driven test automation framework that supports both web and mobile application testing. Its human-readable syntax makes it accessible to testers and developers alike. Robot Framework has extensive libraries and integrations, making it a valuable choice for various testing needs.

4. Appium:

Appium is an open-source automation framework for mobile applications, both Android and iOS. It allows you to write tests using various programming languages, including Python, and provides a unified API for testing mobile apps across different platforms.

5. Behave:

Behave is a Python-based framework for behavior-driven development (BDD). It helps in writing tests in a natural language style, making them more accessible to non-technical stakeholders. Behave encourages collaboration between developers, testers, and product owners.

6. TestNG:

TestNG is a testing framework inspired by JUnit and NUnit, primarily used for Java applications. Although it's not Python-specific, it's worth mentioning for Java-based projects. TestNG supports parallel test execution, data-driven testing, and test configuration management.

7. JUnit:

JUnit is a widely used testing framework for Java applications. Like TestNG, it's not Python-specific but is essential for Java developers. It's known for its simplicity and integration with popular IDEs like Eclipse and IntelliJ IDEA.

8. Allure Framework:

Allure Framework is a reporting framework that can work with various test automation tools, including Selenium and PyTest. It generates interactive and visually appealing reports that provide detailed insights into test execution results. Incorporating these best practices and leveraging these testing frameworks can significantly enhance your test automation efforts. Remember that test automation is an ongoing process, and continuous improvement is key to maintaining the effectiveness of your tests. By adhering to best practices and choosing the right tools, you'll be well-equipped to ensure the quality and reliability of your software throughout its lifecycle.

Chapter 10: Expert-Level Automation Projects and Case Studies

In the realm of automation, we've journeyed through a multitude of topics, from the basics of scripting to advanced concepts of artificial intelligence and machine learning. Now, let's embark on a captivating exploration of real-world automation cases that exemplify expert-level mastery of these principles and techniques.

1. Automated Trading Systems:

Imagine a financial expert who has designed a sophisticated automated trading system that continuously monitors financial markets, executes trades based on complex algorithms, and adjusts strategies in real-time. This expert leverages Python's data analysis libraries and machine learning models to predict market trends and optimize trading decisions, showcasing the power of automation in the world of finance.

2. Autonomous Drone Navigation:

In the world of robotics, an expert has developed an autonomous drone capable of navigating complex environments without human intervention. Through a combination of computer vision, deep learning, and sensor fusion, the drone can avoid obstacles, plan optimal routes, and complete missions such as search and rescue operations or precision agriculture tasks.

3. Healthcare Data Analytics:

Healthcare professionals harness the capabilities of automation to analyze vast amounts of patient data. Expert-level automation techniques, using Python's data processing libraries, allow them to identify disease patterns, predict patient outcomes, and optimize treatment plans. This real-world application demonstrates the immense potential for automation in improving healthcare.

4. Natural Language Processing in Customer Support:

In the customer support industry, automation has reached expert levels with the deployment of natural language processing (NLP) algorithms. Companies employ chatbots and virtual assistants to provide instant responses to customer queries, offering personalized solutions and resolving issues efficiently. These systems continually learn from interactions, showcasing automation's role in enhancing customer service.

5. Self-Driving Cars:

In the automotive sector, experts work tirelessly to perfect self-driving car technology. These vehicles use a combination of sensors, machine learning models, and real-time data analysis to navigate roads, make decisions, and ensure passenger safety. The development of self-driving cars exemplifies automation's profound impact on transportation.

6. Automated Game Testing:

In the gaming industry, automation is instrumental in ensuring game quality. Expert-level automation engineers create test scripts that simulate player actions, uncovering bugs and glitches across different platforms and scenarios. This meticulous testing process guarantees a seamless gaming experience for players worldwide.

7. Predictive Maintenance in Manufacturing:

Manufacturers employ predictive maintenance techniques to prevent equipment failures and production downtime. Using data collected from sensors and IoT devices, experts build machine learning models that predict when machines require maintenance. This proactive approach minimizes disruptions and maximizes efficiency in manufacturing plants.

8. Cybersecurity Threat Detection:

In the realm of cybersecurity, experts leverage automation to detect and respond to threats in real-time. Advanced security systems use machine learning algorithms to analyze network traffic patterns, identify anomalies, and thwart cyberattacks before they can cause harm. This level of automation is crucial in safeguarding sensitive information.

9. Content Generation with AI:
Content creators and marketers harness the power of AI to automate content generation. Expert-level automation systems can generate articles, product descriptions, and marketing copy using natural language generation models. These systems save time and resources while maintaining quality and consistency.

10. Climate Modeling and Prediction:
Climate scientists and environmental experts employ automation to develop intricate climate models. These models simulate complex interactions in the Earth's climate system, helping predict weather patterns, assess the impact of climate change, and inform policy decisions. Automation plays a pivotal role in advancing our understanding of the environment.

11. Space Exploration and Robotics:
In the realm of space exploration, experts deploy robots and rovers to distant planets and celestial bodies. These robots are equipped with advanced automation capabilities, allowing them to perform tasks like collecting soil samples, analyzing rock formations, and sending valuable data back to Earth. This automation enables us to explore the cosmos like never before.

12. Smart Home Automation:
In everyday life, smart home automation experts create systems that seamlessly control lighting, temperature, security, and entertainment. These systems use AI and machine learning to adapt to occupants' preferences, enhancing comfort, convenience, and energy efficiency.

These real-world expert-level automation cases illustrate the incredible breadth and depth of possibilities that automation offers in today's world. Whether it's revolutionizing finance, healthcare, transportation, or entertainment, automation continues to shape industries and drive innovation. As you delve deeper into the world of automation, remember that the only limit to what you can achieve is your imagination and expertise. Automation is a dynamic field that constantly evolves, and the opportunities for creating meaningful solutions are boundless. Imagine an industrial plant where automation rules the roost. Complex automation systems oversee the entire manufacturing process, from raw material handling to quality control and packaging. These systems integrate with various sensors, controllers, and machines, orchestrating a symphony of actions to ensure optimal production efficiency. Now picture an autonomous agricultural ecosystem. Sophisticated automation, driven by sensors, drones, and machine learning models, manages irrigation, pest control, and harvesting. This level of automation not only maximizes crop yields but also reduces the environmental impact of agriculture. In the realm of transportation, complex automation goes far beyond self-driving cars. Imagine an entire smart city where automated traffic management systems dynamically adjust traffic lights and reroute vehicles to optimize traffic flow and reduce congestion. Such systems rely on real-time data analysis, predictive modeling, and communication networks. Space exploration, too, benefits from complex automation. Consider a scenario where a network of autonomous spacecraft and rovers collaborates on an interplanetary mission. Each spacecraft autonomously adjusts its trajectory based on sensor data and communicates with other vehicles to achieve mission objectives. This level of automation enables deep-space exploration and scientific discovery. In the realm of finance, high-frequency trading relies on complex automation. Algorithms analyze market data at lightning speed and execute trades in milliseconds, capitalizing on minute price differences. These algorithms leverage machine learning to adapt to changing market conditions, making split-second decisions with precision.

Advanced automation also plays a crucial role in environmental monitoring and conservation. Autonomous drones equipped with sensors and cameras fly over vast areas of wilderness, collecting data on wildlife populations and habitat health. Machine learning models analyze this data to inform conservation efforts and protect endangered species.

In the healthcare sector, complex automation enhances patient care. Imagine a hospital where robotic systems handle routine tasks such as medication delivery and patient transport. These robots navigate hospital corridors, interact with patients, and ensure that healthcare professionals can focus on critical medical tasks. Complex automation projects are also transforming the world of energy. Smart grids leverage automation to balance electricity supply and demand efficiently. Sensors monitor electricity usage in real-time, allowing for predictive maintenance of power infrastructure and integration of renewable energy sources. Automation is making waves in the world of e-commerce. In large-scale warehouses, fleets of robots collaborate to pick, pack, and ship orders. These robots communicate with each other and with central control systems to optimize the order fulfillment process, ensuring that customers receive their products quickly and accurately. Furthermore, imagine a world where complex automation systems assist with disaster response. Autonomous drones equipped with sensors and cameras can quickly assess disaster-stricken areas, identifying survivors and hazards. Machine learning algorithms process this data, enabling rapid deployment of resources and saving lives. Now, let's talk about the future of automation. As technology continues to advance, we can expect even more complex automation projects to emerge. The fusion of automation with artificial intelligence, quantum computing, and advanced robotics will lead to groundbreaking innovations. Quantum computing, with its immense processing power, will enable complex simulations and data analysis that were previously impossible. This will revolutionize fields such as drug discovery, climate modeling, and materials science.

Advanced robotics will give rise to autonomous systems that can perform tasks in extreme environments, from deep-sea exploration to space missions. These robots will adapt to unforeseen challenges and work alongside humans in collaborative settings. Artificial intelligence will continue to evolve, creating systems that can understand natural language, reason, and learn from vast datasets. AI-driven automation will have applications in fields as diverse as healthcare, legal research, and creative content generation.

The Internet of Things (IoT) will expand the scope of automation by connecting everyday objects and devices to the internet. This will enable smarter homes, cities, and industries, where automation systems respond to real-time data from interconnected sensors and devices.

Blockchain technology, known for its security and transparency, will play a role in automating complex processes such as supply chain management and financial transactions. Smart contracts, powered by blockchain, will automatically execute agreements when predefined conditions are met.

Ethical considerations will also become increasingly important in complex automation projects. As automation becomes more integrated into our lives, questions about privacy, security, and algorithmic bias will need to be addressed. Ethical frameworks and regulations will guide the responsible development and deployment of automation technologies.

In summary, complex automation projects and solutions are shaping the future in remarkable ways. From industrial plants to space exploration, healthcare to disaster response, automation is revolutionizing how we live and work. As technology continues to advance, we can expect even more groundbreaking innovations that will enhance our lives and push the boundaries of what is possible. The future of automation is bright, and it promises to bring about a world of unprecedented convenience, efficiency, and discovery.

Conclusion

In this comprehensive book bundle, "Python Automation Mastery: From Novice to Pro," we've embarked on an incredible journey through the world of automation using Python. Across four distinct volumes, we've explored the entire spectrum of automation, from the foundational principles to advanced strategies and expert-level solutions.

In Book 1, "Python Automation Mastery: A Beginner's Guide," we laid a strong foundation for automation, ensuring that even newcomers to Python could grasp the essential concepts. We started with the basics of Python programming, introduced fundamental automation techniques, and guided readers through real-world examples. This initial step was crucial in building a solid understanding of automation principles. In Book 2, "Python Automation Mastery: Intermediate Techniques," we delved deeper into automation by introducing intermediate-level techniques. We covered topics such as web scraping, scripting, error handling, and more, equipping readers with the skills to tackle complex automation challenges. This volume bridged the gap between beginners and automation enthusiasts looking to enhance their capabilities. Book 3, "Python Automation Mastery: Advanced Strategies," elevated our automation journey to new heights. We explored advanced concepts like data manipulation, object-oriented programming, and leveraging external libraries. Readers learned how to design and implement automation projects with greater sophistication and precision. In the final volume, Book 4, "Python Automation Mastery: Expert-Level Solutions," we reached the pinnacle of automation mastery. Expert-level solutions pushed the boundaries of what's possible, covering complex automation projects in domains ranging from artificial intelligence to network security. By delving into real-world, high-level use cases, readers gained valuable insights into the future of automation. Throughout this book bundle, we emphasized not only the "how" but also the "why" of automation. We explored ethical considerations, best practices, and cutting-edge technologies that shape the landscape of automation. Our goal was to empower readers not just as proficient Python programmers but as automation architects capable of creating innovative solutions. As we conclude this journey from novice to pro in the realm of Python automation, we hope you've not only acquired new skills and knowledge but also developed a profound appreciation for the power of automation in our ever-evolving digital world. Whether you're an aspiring programmer, a seasoned developer, or a curious learner, "Python Automation Mastery" equips you with the tools and insights needed to master the art of automation and take your Python skills to the next level.

Remember that automation is not just a technical skill; it's a mindset that fosters efficiency, productivity, and creativity. As you continue your automation journey beyond these pages, we encourage you to explore, experiment, and innovate. Your newfound mastery of Python automation has the potential to transform industries, solve complex problems, and make a meaningful impact in our increasingly automated society. So, with newfound skills, enthusiasm, and a sense of accomplishment, we invite you to go forth and automate—to harness the limitless possibilities of Python in your pursuit of excellence. The world of automation awaits, and you are now well-prepared to conquer it. Thank you for joining us on this remarkable journey, and may your path be filled with endless opportunities for automation mastery.

About Rob Botwright

Rob Botwright is a seasoned IT professional with a passion for technology and a career spanning over two decades. His journey into the world of information technology began with an insatiable curiosity about computers and a desire to unravel their inner workings. With a relentless drive for knowledge, he has honed his skills and expertise, becoming a respected figure in the IT industry.

Rob's fascination with technology started at a young age when he disassembled his first computer to understand how it operated. This early curiosity led him to pursue a formal education in computer science, where he delved deep into the intricacies of software development, network architecture, and cybersecurity. Throughout his academic journey, Rob consistently demonstrated an exceptional aptitude for problem-solving and innovation.

After completing his formal education, Rob embarked on a professional career that would see him working with some of the most renowned tech companies in the world. He has held various roles in IT, from software engineer to network administrator, and has been instrumental in implementing cutting-edge solutions that have streamlined operations and enhanced security for businesses of all sizes.

Rob's contributions to the IT community extend beyond his work in the corporate sector. He is a prolific writer and has authored numerous articles, blogs, and whitepapers on emerging technologies, cybersecurity best practices, and the ever-evolving landscape of information technology. His ability to distill complex technical concepts into easily understandable insights has earned him a dedicated following of readers eager to stay at the forefront of IT trends.

In addition to his writing, Rob is a sought-after speaker at industry conferences and seminars, where he shares his expertise and experiences with fellow IT professionals. He is known for his engaging and informative presentations, which inspire others to embrace innovation and adapt to the rapidly changing IT landscape.

Beyond the world of technology, Rob is a dedicated mentor who is passionate about nurturing the next generation of IT talent. He believes in the power of education and actively participates in initiatives aimed at bridging the digital divide, ensuring that young minds have access to the tools and knowledge needed to thrive in the digital age.

When he's not immersed in the realm of IT, Rob enjoys exploring the great outdoors, where he finds inspiration and balance. Whether he's hiking through rugged terrain or embarking on a new adventure, he approaches life with the same inquisitiveness and determination that have driven his success in the world of technology.

Rob Botwright's journey through the ever-evolving landscape of information technology is a testament to his unwavering commitment to innovation, education, and the pursuit of excellence. His passion for technology and dedication to sharing his knowledge have made him a respected authority in the field and a source of inspiration for IT professionals around the world.

www.ingramcontent.com/pod-product-compliance
Lightning Source LLC
Chambersburg PA
CBHW071236050326
40690CB00011B/2137